NEW ENGINEERING TECHNOLOGIES FOR MODERN FARMING

NEW INDIA PUBLISHING AGENCY
New Delhi-110 034

About the Authors

Prof. Narendra Singh Rathore is Vice Chancellor, Maharana Pratap University of Agriculture and Technology, Udaipur. He did M. Tech with specialization in Energy Studies from I.I.T., Delhi and Ph.D. from Rajasthan Agricultural University, Bikaner. He has served the Indian Council of Agricultural Research (ICAR) as Deputy Director General (Engg.), and Deputy Director General (Education). He was also founder Vice Chancellor of Sri Karan Narendra Agriculture University, Jobner- Jaipur and also worked as Vice Chancellor in Mohanlal Sukhadia University, Udaipur. He has contributed more than 200 technical papers and 36 books on various aspects of Renewable Energy, Environment and Agricultural Engineering. He has undertaken 51 research projects and organized 20 International trainings, Seminars, Workshops and Conferences and 12 ISTE sponsored Summer School etc. There are more than 100 reports, proceedings and popular articles to his credit. He was honored by 31 National & International Awards from governmental, society and private organization.

Dr. N. L. Panwar is an Assistant Professor of Renewable Energy Engineering in the Faculty of Engineering at MPUAT, Udaipur, Rajasthan. He has been awarded PhD from the Centre for Energy Studies, Indian Institute of Technology Delhi. His areas of interest cover energy and exergy analyses of thermal systems and energy efficiency and management. He is actively involved in design and development of improved cook stoves, biomass gasifier, and solar thermal devices for industrial and rural applications. He has contributed more than 100 papers in international journals and several books on Renewable Energy aspects. He has been a recipient of 'Prakritik Urja Puraskar' from Ministry of New and Renewable Energy, Government of India for his outstanding book on Alternative Energy Resources. He has also been awarded by "Shrimati Vijay-Usha Sodha Research Award" from Indian Institute of Technology, Delhi in 2014, and Rajasthan Energy Conservation Award 2018 from Government of Rajasthan.

NEW ENGINEERING TECHNOLOGIES FOR MODERN FARMING

Narendra Singh Rathore
Vice Chancellor
Maharana Pratap University of Agriculture and Technology
Udaipur (Rajasthan) India

N.L. Panwar
Department of Renewable Energy Engineering
Maharana Pratap University of Agriculture and Technology
Udaipur (Rajasthan) India

NEW INDIA PUBLISHING AGENCY
New Delhi-110 034

NEW INDIA PUBLISHING AGENCY
101, Vikas Surya Plaza, CU Block, LSC Market
Pitam Pura, New Delhi – 110 034, India
Email: info@nipabooks.com
Web: www.nipabooks.com

For customer assistance, please contact
Phone: + 91-11-27 34 17 17 Fax: + 91-11- 27 34 16 16
E-Mail: feedbacks@nipabooks.com

ISBN: 978-93-90591-31-2

Composed and Designed by NIPA.

Preface

The global populations is continuously increasing in spite of all types of family planning programme and polices and will continue for the next half century. Therefore, more food and fibre will be required to feed the growing population. The agriculture is an important source of survival for all living beings because it produces food, feed, fibre, fertilizer and fuel. In the present context, modern agriculture evolving various innovative farming techniques improves farmers' income, increase efficiency, help in conservation of natural resources such as water, soil, and energy etc. To secure future of humankind, more emphasis is given to soil and water engineering. Soil and water will play a key role to produce sufficient food for growing population in future. Advancement in the field of science and technology is the driving force of advance research in the field of Indian Agriculture, which will enable the Indian farmers to feed the growing population. This book is written with aim to understand the engineering inputs in modern agriculture for improving production, productivity and profitability.

The book is a ready reference material for students from a wide variety of engineering and science back grounds wish to study the Engineering Aspects of Agriculture Engineering.

This book has been divided into ten chapters, which cover new dimensions of Engineering in Agriculture. **Chapter 1** deals with an overview of Indian agriculture, engineering intervention in agriculture, and concept of sustainable agriculture. **Chapter 2** pertains to the application of protected cultivation for better yield. Different types of greenhouses and thermal modeling of naturally ventilated greenhouse have also been discussed in this chapter. **Chapter 3** brings out the cultivation practices in vertical farming, types of vertical farming, and renewable energy applications in vertical farming practices. **Chapter 4** deals with precision agriculture which is an advance farm management with the help of information and communication / digital technology to confirm that the cultivated crops and soil receive desire inputs for optimum health and productivity. **Chapter 5** brings out the crop production under hydroponics conditions. The combinations of aquaculture and hydroponics plant production in a close-loopwater system where both vegetables and fish grow simultaneously are discussed in **Chapter 6**

Chapter 7 deals with secondary agriculture and value addition of primary agricultural produce. The artificial controlled environmental conditions for growing plants are discussed in **Chapter 8** under phytotrons technologies. Applications of Artificial Intelligence and Robotics in Agriculture are discussed in **Chapter 9 and 10** respectively. It is hoped that this book will be useful as text book and as reference book for students pursuing course in agricultural sciences and engineering studies. This will also give new dimensions to research scientists for their critical thinking and dissemination of technologies.

Authors

Contents

1

Introduction

1.1 General

Food is basic necessity for the survival of human beings. It is derived from agriculture which traps and transfers solar energy in chemical energy in the form of biomass and end products are processed and utilized for food, feed, fibre, fertilizer, and fuel. Indian's economic security continues to be foreseen upon agricultural sector. Agriculture supports 52 per cent of population, as against about 75 per cent at the time of dependence. Despite the focus on industrialization, agriculture remains a dominant sector of the Indian economy both in terms of contribution to Gross Domestic Product (GDP) as well as a source of employment to millions across the country. India occupies a leading position in global trade of agricultural products. However, its total agricultural export basket accounts for a little over 2.15 per cent of the world agricultural trade. The major export destinations are USA, Saudi Arabia, Iran, Nepal and Bangladesh. Among agricultural commodities, basmati rice, spices, oil meals, sugar, cotton and castor oil have been leading export commodities[1].

Agriculture is an essential ingredient in overall development of India economic. The history of Indian agriculture in dates back to Indus Valley Civilization and even before that in some places of Southern India. India ranks second worldwide in farm outputs.

On the eve of Independence, 70 per cent of India's nation income was from agricultural sector. Before 1947, over 85 per cent of country's population lives in villages where livelihood completely depended on agriculture, which contributed over 95per cent of country's income. Agricultural sector was most important sector in Indian economy facing massive stagnation and continuous determination. In the present context, around 60 – 70 per cent of Indian population directly or indirectly depends upon the agricultural sector and currently it contributes to 16 per cent of the Indian gross domestic product.

Agriculture in India played major role even in an outbreak of novel coronavirus. As per the survey report of Confederation of Indian Industry (CII), about 45 per cent of CEOs in India said they don't see economic normalcy returning before a year. Another 36 per cent were more optimistic but said it would take 6-12 months for economy to function with normality. The Indian economy of industry and services sectors (agriculture fields don't have CEOs) is to reel under the impact of coronavirus. In such a pandemic situation, Indian economy is left with agriculture and only agriculture to depend upon. And, the good news is India is expecting record, food-grain production by next year at almost 300 million tonnes -- 298.32 million tonnes to be precise (149.92 MT kharif + 148.4 MT rabi). It is important to notice that, in India agriculture provides employment to about 55 per cent of workforce in India[2]. Therefore, it is essential to make reform to uplift the agricultural sector to raise the income of population dependent on agriculture and growth of non-agricultural sector. Reform in the agriculture sector should incorporate an efficient delivery systems of public service, social mobilization, and community participation. Apart from a said point, there are three possible goals in development of agricultural sector to be considered: (a) achieving high growth by raising productivity; (b) inclusiveness by focusing on lagging regions, small farmers and women; and (c) sustainability of agriculture[3].

India has remained at the lower end of the global agricultural export value chain given that majority of its exports are low value, raw or semi-processed and marketed in bulk. The share of India's high value and value added agricultural produce in its agricultural export basket is less than 15per cent compared to 25per cent in US and 49per cent in China. India is unable to export its vast horticultural produce due to lack of uniformity in quality, standardization and its inability to curtail losses across the value chain. Given the globalization of value chains, it is imperative that the country makes concerted efforts to boost exports of high margin, value added and branded processed products. The policy will involve a paradigm shift from residual export after meeting domestic demand to targeted export according to preferences of overseas market.

1.2 Information Technology and Agriculture

This is the era of Information Technology, as it plays major role in enlightening the path of sustainable development. Global liberalization and interdependent global economy, morden technology has emerged as the driving force behind all kinds of production and trade. In broader prospect, technology is considered a tool to transforms the natural resources into different goods and services extensively used by society[4]. All kinds of natural energy sources can be conserved by improving system / process efficiency, information technology plays vital roles in the energy efficiency.

Information technology is nothing, but it is the application of technology to solve business or organizational problem on a large scale with the use of any computers, storage, networking and other physical devices, infrastructure and processes to create, process, store, secure and exchange all forms of electronic data. In other words, Information Technology (IT) is a broad generic term for the electronic storage, processing and subsequent retrieval of data to demonstrate the relationships between data sets. The word data can be defined as the "numerical representation or other symbolic surrogates aiming at characterizing attributes of people, organizations, objects, events or concepts[5].

A case study on effects of process change during a period of 12 years and impact of information technology on energy efficiency in steel reheating furnaces was conducted by Mårtensson[6]. Based on the regression model, it was concluded that the improvement in energy efficiency was largely facilitated by the new control system. Further, information technology has been shown to supply possibilities of enhancing the operation of an industrial process in terms of improved energy efficiency[6].

Information technology in the agricultural sector directly contributes in its productivity and also empowering farmers to take informed and quality decisions, which will have a positive impact on the way agriculture and allied activities are conducted. Extensive uses of IT in precision farming shall contribute in better agricultural productivity. Technological evolutions in remote sensing, Geographical Information Systems (GIS), soil sciences, and agronomy are also directly contributed in overall agricultural productivity. It is empowering the formers by providing the timely and reliable source of information for taking timely decisions. Information technology in agriculture helps in improving timely action and making right decision, better planning, and adaptation of better farming and harvesting methods.

In Indian context, lack of awareness about the advancement in technologies, the potentials of IT in agricultural sectors are still unexploited. Indian farmers are still hesitating to come out of the entanglements of traditional source of inputs. Using information is not only useful but a requirement in these days. Studies showed that the use of modern technologies is capable of making remarkable hike in agricultural production[7].

Information technology plays major role in the Indian education system. When India gained independence in 1947, the nation had a total of 241,369 students registered across 20 universities and 496 colleges. As of 2020, India has about 1000 universities, with the breakup of 50 central universities, 402 state universities, 125 deemed universities. 334 private universities, 155 Institutes of National Importance which include IITs, AIIMS, IIMs, IISERs, NITs, and 30.9 million students as reported by MHRD in 2020. India's higher education system is the third largest in the world, next to the United States and China. As far as agricultural education is concerned, the Indian Council of Agricultural Research (ICAR) is the main regulatory authority of agricultural education in India. There are three Central Agricultural Universities (CAUs), four ICAR Deemed Universities, 4 central university with agricultural faculty and 63 State Agricultural Universities (SAUs) as of December 2019. The information technology sharpens the agricultural research, education and extension to improve the life in rural areas and impart updated scientific knowledge among the students.

1.3 Engineering Intervention in Agriculture

Prior to independence, Bengal famine in 1943 during the British rules was one of the most severe famines that killed over 3 million people. During that time industrial sector was not that much developed, and no help was extended from this sector, and undoubtedly agriculture sector was the biggest disaster for the peoples of India during British regime. After independence, India realized the need of self-sufficient in food production to feed the people even is worst situation. In 1965, Indian scientist Dr. M.S. Swaminathan found Green Revolution which began with adoption of high yielding and disease resistant wheat varieties with advance farming knowledge to increase the productivity. Enabled the country to overcome from hunger and starvation, to achieve self-sufficiency in food, reduce poverty, and improve the living standard of millions of rural farmers and family. During green revolution period, Indian agricultural practices shifted from conventional practices to more technological practices. During green revolution, more emphasis was given of high yield varieties (HYVs) of plants and grains. Advance cultivation practices spread in Punjab, Haryana, Western Uttar Pradesh, Tamil Nadu, Kerala, etc. The technique used during green revolution has caused and continues to cause irreversible changes, including deterioration in soil health and nutrients. This practice makes it difficult to shift back to organic farming on the same land. In the years since its independence, India has made immense progress towards food security. Indian population has tripled, and food-grain production more than quadrupled. There has been a substantial increase in available food-grain per capita.

1.3.1 Sustainable Agriculture

It is bitter truth that the majority of foods available in the market are produced using industrialized agriculture. Industrialized agriculture is nothing but it is a type of agriculture where large quantities of crops and livestock are produced through industrial techniques for the purpose of sale. This type of agriculture mainly relies on a variety of chemicals and artificial enhancements, such as pesticides, fertilizers, and genetically modified organisms. Industrialized agriculture has made it possible to produce large quantity of food to feed 1.35 billion Indian peoples. It has the negative aspect at large. Therefore, there is a need to shift towards sustainable agriculture.

Sustainable agriculture is a technique of agriculture, which mainly aimed to produce crops and livestock with least effects on environment. In this technique, attempt has been made to find a good balance between the need for food production and the preservation of the ecological system within the environment. Effective water utilization and conservation, reducing the use of fertilizers and pesticides, and promoting biodiversity in crops grown and the ecosystems are the basis goal associated with sustainable agriculture. In addition to this, it has also focused on maintaining economic stability of farms and helping farmers improve their techniques and quality of life. Sustainable agriculture contributes a lot to environment, almost 30 per cent less energy is being used per unit crop yield as compared to industrialized agriculture. Dependencies on fossil fuel, pesticide, and insecticide are significantly reduced. It also contributes in increasing the biodiversity of the area by providing a variety of organisms with healthy and natural environments to live in[8]. Agriculture is sustainable when it cultivation practices are ecologically sound, economically viable, and socially acceptable at large. As Shri Narendra Modi, Hon'ble Prime Minister, Government of India has targeted doubling the income of Indian farmers by 2022, indeed it is an ambitious target no doubt, it is certainly achievable if right tools and measures are adopted during farming.

1.3.2 Farm Mechanisation

The production of food grains and cash crop in India has increased significantly. However, it has been observed that there is stagnation in the per unit productivity due to constraints like unavailability of farm manpower, urbanisation, migration of farm labourers to the non-farm sector, and reduced livestock rearing on farms. This is the main reason that income of Indian farmers' is unable to maintain the pace. Production cost is reasonably increasing over the periods. Therefore, there is an urgent need to identify the labour substitution. In this respect farm mechanisation is the only option which can cut input cost as a whole. As per the World Bank estimates, half of the Indian population would be urbanised by the year 2050. There will be sharp drop of agricultural workers in total work force to 25.7 per cent by 2050 from 58.2 per cent in 2001. Thus, there is necessity to boost the level of farm mechanization in the country. Farm mechanization helps in providing optimal utilization of resources, save valuable time and reduces drudgery of farmer engaged in manual operations. Small and marginal farmers are unable purchase farm implemented; this is the one of the reason for slow uptake of farm mechanisation. There is no doubt that effective use farm implementes and agricultural machinery helps in increasing the productivity and production, timely action of farm operation and enable farmers to quickly rotate the crops on the same land. High level of farm mechanisation has been observed in Indian states such as Punjab, Haryana, and UP. However, negligibles mechanization seen in north-eastern states. The level of farm mechanization in India stands at about 40-45 per cent to that of global level. This level of farm mechanization is still low as compared to the countries such as the U. S. (95 per cent), Brazil (75 per cent) and China (57 per cent). While the level of mechanization lags behind other developed countries, it has seen an average agriculture growth rate of 3.56 per cent through the last decade[9].

With the shrinking land and water resources and labour force, the onus rests on mechanization of production and post harvesting operations. There is a linear relationship between availability of farm power and farm yield and Government has decided to enhance farm power availability from 2.02 kW per ha (2016-17) to 4.0 kW per ha by the end of 2030 to cope up with increasing demand for food grains. Recognising the need for inclusive growth of farm mechanization sector in the country, a sub mission on Agricultural Mechanization was launched in 2014-15. Under the scheme, assistance is provided to State governments to impart training and demonstration of agricultural machinery. It provides assistance to farmers for procurement of various agricultural machineries and equipment and for setting up of Custom Hiring Centre[10]. Mechanization in agricultural allied sectors such as dairy and livestock is largely increased, i.e. milking machines, fodder handling and harvesting systems is few examples. The key growth drivers of farm mechanisation in India are categorised as social, agricultural and agronomical and economic factors as illustrated in Fig. 1.1.

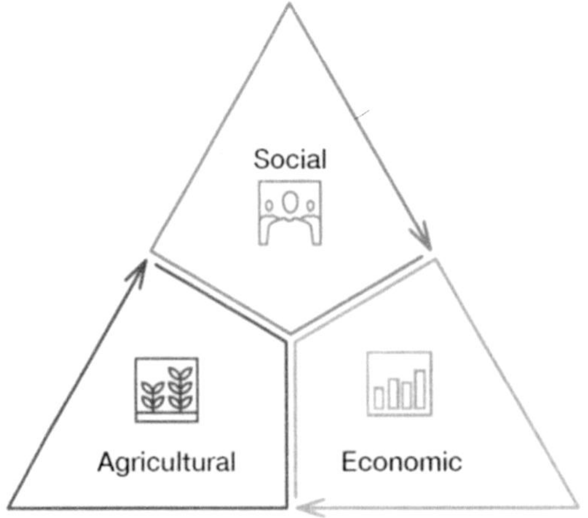

Fig. 1.1: Key Drivers of Farm Mechanisation in India

Under the social driver, it has been generally observed that migration of men is increasing for better prospects in non-farm opportunities has rendered farming a responsibility for women, especially in poorer states. The physically demanding work such as weeding operation, which is more difficult for women to perform and increase the drudgery. Female labourers across the country actively involved in activities such as paddy transplanting, dibbling and harvesting. These activities are not only labour-intensive but also back-breaking, makes mechanisation a necessity.

To enhance the income per unit of land, more crops are being cultivated by farmers. Certainly, time available between two crops reduces, and it becomes more important to carry out agricultural operation in time. Farmers are vulnerable to extreme climatic conditions and vagaries of weather; in such a situation, mechanised farm operations help farmers mitigate/adapt to climate risk by carrying out farming operations in a short duration or suitable window. Undoubtedly, mechanising operations resultes in time and cost savings, thus making farming activity more profitable[11].

1.3.3 Renewable Energy Technologies in Agriculture

Renewable technologies are considered to be clean sources of energy. The optimal use of these resources minimize environmental impacts, produce minimum secondary wastes, and are sustainable based on current and future economic and social needs. Sustainable development requires methods and tools to measure and compare human activities' and its environmental impacts for various products. Renewable Energy Sources (RES), which include biomass, hydropower, geothermal, solar, wind, and marine energies, supply 14 per cent of the world's total energy demand. Renewable resources refer to primary, domestic, and clean or inexhaustible energy resources[2]. Large-scale hydropower energy supplies 20 per cent of the global electricity need. Wind power in coastal and other windy regions is a promising source of energy.

Renewable energy resources can make a decisive contribution to the economic, social and sustainable development of rural regions in developing countries, yet the consumption of fossil fuels is dramatically increasing along with improvements in the quality of life, the industrialization of developing nations, and the increase of the world's population. It has long been recognized that excessive consumption of fossil fuel leads not only to a diminishing fossil fuel reserves more quickly, but also has a significant adverse impact on the environment. Such impacts result in increased health risks and the threat of global climate change. Changes to improve environmental conditions are becoming more politically acceptable globally, especially in developed countries. Society is slowly moving towards seeking more sustainable production methods, minimizing waste, reducing air pollution from vehicles, generating distributed energy, conserving native forests, and reducing greenhouse gas emissions. Increasing consumption of fossil fuel to meet out current energy demands, however, has sounded alarms over regarding potential energy crisis. This has generated a resurgence of interest in promoting renewable alternatives to meet the developing world's growing energy needs. Excessive use of fossil fuels has caused global warming from carbon dioxide emissions, therefore, promoting renewable forms of clean energy. To monitor the level of these greenhouse gas emissions, an agreement that has fulfiled the objectives of the Kyoto Protocol was created with overall pollution prevention targets.

Renewable energy and agriculture farming are an attractive combination. Solar, biomass, and wind can be harvested forever, providing farmers with a long-term source of income. Renewable energy can be used on the farm to replace other fuels or sold as a "cash crop." The amount of energy from the sun that reaches Earth each day is enormous. All the energy stored in Earth's reserves of coal, oil, and natural gas is equal to the energy from only 20 days of sunshine. While desert areas Rajasthan and western part of Gujarat get more sun than other parts of the country, most Indian states receive enough sunshine to make solar energy practical. Solar energy can be used in agriculture in a number of ways for saving money, increasing self-reliance, and reducing pollution. Solar energy can cut a farm's electricity bills. Solar heat collectors can be used to dry crops, and cultivation of crop in greenhouses. Solar water heaters can provide hot water for dairy operations, pen cleaning applications. and homes. Photovoltaics (solar electric panels) can power farm operations and remote water pumps, lights, and electric fences. Buildings and barns can be renovated to capture natural daylight, instead of using electric lights. Solar power is often less expensive than extending power lines. Hence, to reduce the greenhouse-gas emissions in the agricultural sector, there is a need to shift conventional energy generation into renewable-energy generation. Renewable energy contributes a lot in different agriculture unit operation right from land preparation to finished products as illustrated in Fig. 1.2.

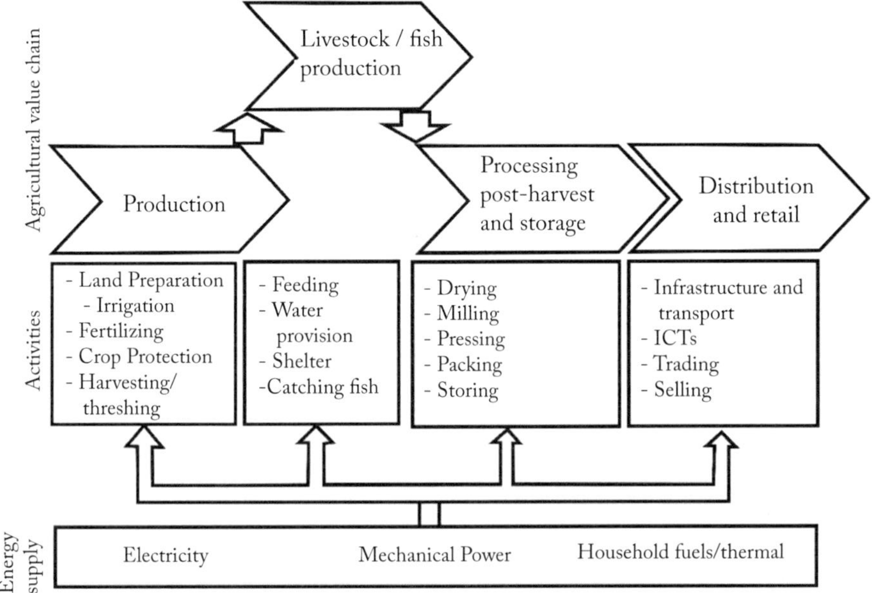

Fig. 1.2: Energy Involvement in Different Agricultural Unit Operations

Biomass energy is produced from plants and organic wastes - everything from crops, trees, and crop residues to manure. Crops grown for energy could be produced in large quantities, just as food crops are. Crops and biomass wastes can be converted to energy on the farm. Crop residues can be densified to produce industrial fuel. Pyrolysis techniques help to achieve pyrolytic oil and biochar from crop residues. Motive power can be generated by combustion of producer gas generated during gasification of biomass. Further, producer gas can be effectively utilized to generate process heat for drying the crop at large. Wind energy also significantly contributes in Indian agriculture. Wind power is extensively used to pump water for irrigation and to generate electricity[12].

1.3.4 Natural Resource Management

Natural Resource Management deals with managing the way in which people and natural landscapes interact. It brings together land use planning, water management, bio-diversity conservation, and the future sustainability of agriculture. Agriculture is considered as a natural resource because it, requires fertile soil, with nutrients. Soil is a natural resource that provides, minerals and water to plants. Soil is one of the most important and essential natural resources. Farmers understand that when the soil is destroyed, good crops cannot grow. Soils can effectively remove impurities, kill disease agents, and degrade contaminants. Typically, soils maintain a net absorption of oxygen and methane, and undergo a net release of carbon dioxide and nitrous oxide. Soils offer plants physical support, air, water, temperature moderation, nutrients and protection from toxins. Soils provide readily available nutrients to plants and animals by converting dead organic matter into various nutrient forms. It is made up of air, water, minerals and organic material and is one of the most important natural resources on earth. Most life on earth depends on soil as a direct or indirect source of food. Plants and animals source their nutrients from the soil and it is home to many different forms of life. On the other hand, it is essential to maintain balance in ecosystem. Productive capacity of farm reduced with loss of ecosystem, resulting lower the income of farmers. Vermicomposting is a low-technology, environmentally-friendly useful conversion process of rural and urban organic waste into good quality manure. The resulting vermicompost has been shown to have several positive impacts on plant growth and health. This organic fertilizer is increasingly considered in agriculture and horticulture as a promising alternative to inorganic fertilizers and/or peat in greenhouse potting media[13]. Its nutritional quality is better and higher contents of essential plant nutrients and numerous growth promoters. Further, it improves physical, chemical and biological environment of soils and overall crop productivity. Apart from above mentioned advantages, it provides employment to rural households[14].

References

1. Tiwari R. Economic Survey suggests to boost agri exports. The Economic Times. https://economictimes.indiatimes.com/news/economy/agriculture/economic-survey-suggests-to-boost-agri-exports/articleshow/73800027.cms accessed on July 17, 2020.
2. Dutta PK. Indian economy in corona time: Agriculture only bright spot. https://www.indiatoday.in/news-analysis/story/coronavirus-lockdown-covid-19-impact-on-economy-agriculture-1674545-2020-05-05 accessed on July 15, 2020
3. Dev SM. Transformation of Indian Agriculture? http://www.indiaenviron mentportal.org.in/files/file/TRANSFORMATION%20OF%20INDIAN %20AGRICULTURE.pdf accessed on July 15, 2020
4. Balachandra P, Nathan HS, Reddy BS. Commercialization of sustainable energy technologies. Renewable Energy. 2010 Aug 1;35(8):1842-51.
5. Moffatt I. The potentialities and problems associated with applying information technology to environmental management. Journal of Environmental Management. 1990 Apr 1;30(3):209-20.
6. Mårtensson A. The impact of information technology on energy efficiency: an evaluation case study of steel-reheating furnaces. Energy. 1994 Jul 1;19(7):717-28.
7. Web page: https://orisys.in/role-of-information-technology-in-agriculture/# accessed on July 14, 2020
8. Web page: What Is Sustainable Agriculture? https://study.com/academy/ accessed on July 15, 2020.
9. NABARD. Sectoral Paper on Farm Mechanization. https://www.nabard.org/auth/writereaddata /file/NSP%20Farm%20Mechanisation.pdf accessed on July 16, 2020.

10. Tiwari R. Focus on farm mechanization to cope up with increasing food demand.https://economictimes.indiatimes.com/news/economy/agriculture/focus-on-farm-mechanization-to-cope-up-with-increasing-food-demand/articleshow/73814116.cms accessed on July 16, 2020
11. FICCI Farm mechanisation: Ensuring a sustainable rise in farm productivity and income. http://ficci.in/spdocument/23154/Online_Farm-mechanization-ficci.pdf accessed on July 16, 2020
12. REA Renewable Energy and Agriculture: A Natural Fit https://www.ucsusa.org/resources/renewable-energy-and-agriculture accessed on July 16, 2020
13. Lazcano C, Domínguez J. The use of vermicompost in sustainable agriculture: impact on plant growth and soil fertility. Soil nutrients. 2011;10(1-23):187.
14. Ali N. Engineering Interventions for Higher and Sustainable Agricultural Growth In India. http://www.un-csam.org/Activities%20Files/A0711/01in.pdf accessed on July 17, 2020

2

Protected Cultivation

2.1 Introduction

The great impact of increase in population as well as rapid economic growth has given us cause for concern about the future situation of the earth in recent years. Desertification as well as deforestation is continuously changing the landscape. Irrigated areas are being damaged by salinization. The effect to increase arable lands cannot match with the speed of these deteriorating processes. Weather is also being changed gradually by urbanization. In the new millennium, the challenges in the agricultural sector are very different from those met in the previous decades. The enormous pressure to produce more food from less land with shrinking natural resources is a tough task for the farmers. This calls for special effort to manage the key input without eroding the ecological assets and sound knowledge base to sustain agricultural productivity and profitability[1].

The maximum crop response depends on the level of the balanced environmental parameters. The natural environment and input availability may not be optimum for a given crop. The environmental factors which affect plant growth includes air temperature, relative humidity, carbon dioxide concentration soil temperature and moisture content to the soil. The micro-climate can be artificially controlled by means of plastic covering structure like greenhouse. The plastic covered greenhouse is the only mechanism, which transmits the useful wavelengths of the light spectrum for photosynthetic activity. Further, it also reduces the air movement, which is of great importance for the plant growth and often used to control the spread of plant diseases[2].

As far as plant propagation is concerned sunlight provides the primary energy for photosynthesis, the food chain and all human nutrition. The expansion of crop and flower production in various types of greenhouse during the recent years has enabled the growth of agricultural products throughout the year[3,4]. An agricultural greenhouse consists of frames of metallic or wooden structure covered with a transparent material preferably plastic which provide a suitable environment for the intensive production of various crops. A greenhouse is essentially an enclosed structure, which traps the short wavelength solar radiation and convert and finally stores the long wavelength thermal radiation to create a favorable microclimate for higher productivity[5,6]. Open field agricultural farming has no control on the environmental parameters such as sunlight, air composition and temperature that influence the plant escalation. Hence, a large number of winter vegetables, flowers and other horticultural crops have to be transported from distant places[7]. Future greenhouse systems in moderate climates will have to be energy efficient. Greenhouses will have decreased transmission for long-wave length radiation and increased transmission for short-wave-radiation compared with common shelters[8].

Computerised control is an intrinsic part of modern greenhouses. The functions of the climate computer can be summarised as follows: (i) it takes care of maintaining a protected environment despite fluctuations external weather (controller function), (ii) it acts as a programme memory, which can be operated by the grower as a tool to steer his cultivation[9]. Its control is moving towards model-based optimal systems[10], which results in the development of various kinds of dynamic models[11].

Sun is the main sources of energy in the universe. Photosynthesis is primarily needed for plants to convert the CO_2 and water into the food. The CO_2 level in the greenhouse structure is much higher in the morning hours as compared with open field conditions. It enhances the overall productivity of growing crop. The expansion of crop and flower production in various types of greenhouse during the recent years has enabled the growth of agricultural products through the entire year[4]. Protected cultivation has got a very good potential and is also controls focus of the agricultural economy globally. As per the World Greenhouse Vegetable Statistics- 2019 updates, the estimated global protected agriculture area is 5,630,000 ha. (13,912,000 ac.). As a comparison, the published global greenhouse vegetable area in 1980, was 150,000 ha. (371,000 ac.) whereas in 1995, it was 500,000 ha. (1.2 million ac.).

A greenhouse is the quasi-permanent structure, covered with a transparent or translucent material, ranging from simple homemade designs to sophisticated pre-fabricated structures, wherein the environment could be modified suitable for propagation or growing of plants. Materials used to construct a greenhouse frame may be wood, bamboo and steel or even aluminum. Covering can be glass or various rigid or flexible plastic materials. Depending on the covering material, different terminology has been used in the context of greenhouse structures as mentioned below:

Glasshouse: - A greenhouse with glass as the covering material is called glasshouse

Polyhouse: - It is a greenhouse with polyethylene as the covering material.

Net house: - It is covered by net type of cover.

2.2 Plant Environment and Greenhouse Climate

A plant grows best when exposed to an environment that is optimal for that particular plant species. The aerial environment for the plant growth can be specified by the following four factors:

1 Heat or temperature

2 Light

3 Relative humidity

4 Carbon dioxide

While plants have precise optimum environmental conditions for best growth, most are tolerant to variations in these conditions within some limits. However, permanent damage would occur when they are exposed to conditions outside these limits. At the same time plants are subjected to attack by pest and disease.

Greenhouse crop production is protected against adverse environmental conditions and allows pest and disease to be excluded or controlled. Besides providing protected enclosure, a greenhouse also acts as a 'Heat trap'. It admits solar radiation and converts this energy into heat by raising the temperature of the greenhouse air. While this is the basis of the greenhouse's ability to perform its tasks, it also affects others environmental factors.

Environmental conditions inside the greenhouse can be modified suiting to the potential growth of the plants. The extent of climate modification will however, depend on the design of greenhouse and is generally related with its cost. Higher the capacity of greenhouse to modify its climate, higher is the cost of its construction.

The way in which a greenhouse gets heated exposed to sunlight is similar to heating of the earth's surface end its adjacent atmosphere. When solar radiation reaches the earth small portion is reflected back into the space while the remainder is absorbed at the surface raising its temperature. In the same way, when solar radiation reaches to greenhouse cover- surface, a small amount (normally – 15-20%) is reflected back from the surface while the remainder is transmitted to the interior. Plants, soil and other objects absorb most of this transmitted radiation and remainder is reflected as shown in Fig 2.1

The absorbed radiation raises the temperature of absorbing surfaces and objects with the heat energy being immediately transferred to the greenhouse air by convection and evaporation, thereby increasing the temperature and humidity.

Greenhouse loses heat in three ways

1 Warm air moves out of the greenhouse and is replaced by cooler outside air.

2 Heat is transferred through the covering material itself.

3 Heat is lost through the soil/floor

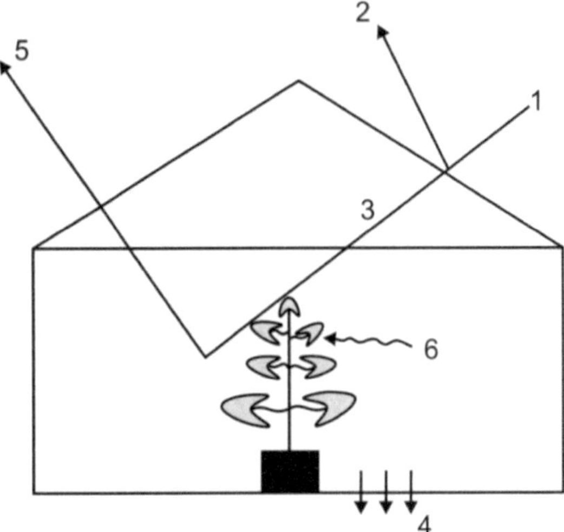

Fig. 2.1: Solar Radiation Heats the Greenhouse

1. Solar Radiation Reaching the Greenhouse, 2. Solar Radiation Reflected from Greenhouse Surface, 4. Solar Radiation Transmitted to Interior of Greenhouse, 5. Absorbed Radiant Heat Energy Conducted into Surface, 6. Solar Radiation Reflected from Internal Surfaces.

Increasing the temperature of the growing environment in a greenhouse is unavoidable, but it is also the most important function of a greenhouse. Enhanced temperatures accelerate and allow sustaining plant growth even when outside ambient temperatures are unfavorably low. However, during summers inside temperatures rise higher than the optimum level and therefore, cooling/ ventilation provision are necessary.

Most plants grow better within 60-80% relative humidity (RH) of air. Low RH increases the evaporative demand on the plant, while high RH can depress this demand inhibiting the uptake of nutrients, particularly calcium. In general, the RH inside the greenhouse is higher than the outside, mainly due to transpiration load. Effective ventilation is required to control higher RH level.

In most part of the country, solar radiation is not a limiting factor for the plant growth. Light control inside the greenhouse can be affected conveniently either by shading or by supplementary lighting whenever required. Growers in northern India should, however be careful in monitoring light level in winters especially during prolonged foggy conditions. As per-urban areas, particulate pollutants get deposited on the plastic roof thereby reducing the light transmission significantly. This problem is compounded in winter. In such conditions, it is necessary to wash the roof frequently to maintained adequate light level inside the greenhouse.

Plants use carbon dioxide from the atmosphere for photosynthesis. Carbon dioxide concentration inside the greenhouse in the early morning is always higher than the outside. With the onset of sun, this level quickly depletes and goes down the normal level during the day if adequate air exchange is not maintained. Carbon dioxide enrichment is generally accomplished by burning suitable fuel like propone.

2.3. Effect of CO_2 Enrichment

It is almost 200 years since the positive effects of CO_2 enrichment on plant growth were first observed. From about 1900 to the early 1930's, extensive CO_2 researches were carried out in different European countries[12-14] and in the U.S.A.[15]. Carbon dioxide enrichment decreases the oxygen inhibition of photosynthesis and increases the net photosynthesis in plants. This is the basis for increased growth rates caused by CO_2 at low as well as at high light levels. Elevated CO_2 concentrations also increase the optimal temperature for growth. Pot plants, cut flowers, vegetables and forest plants show very positive effects from CO_2 enrichment by increased dry weight, plant height, number of leaves and lateral branching. Plant quality expressed by growth habit and number of flowers is often enhanced by CO_2 enrichment. The rooting of cuttings is often stimulated by high CO_2 levels. The optimal CO_2 concentration for growth and yield seems to lie between 700 and 900 ml l^{-1}, and this CO_2 level are generally recommended in greenhouses. Carbon dioxide concentrations higher than 1000 ml l^{-1} might cause growth reductions and leaf injuries, and certainly increases the loss of CO_2 due to leakage from the greenhouse[16].

Raising daytime temperature for crop fertilized with CO_2 has been generally beneficial. While raising the night time temperature has not also recommended an increase as much as 10°F (6°C) temperature for roses[17]. Ioslovich et al.[18] reported that greenhouse CO_2 enrichment, in warm climate is restricted by the need to ventilate, leading some growers to be intermittent where enrichment and ventilation alternate several times an hour. This strategy relies on the heat and CO_2 capacity of the system, characterized by a heating times constant of the order of 10 min, during which weather; the period ventilation may be suspended. It was shown that, for slowly changing weather the optimal CO_2 enrichment is basically not intermittent, but rather is quasi steady state. As the disturbance frequency increase, the quasi steady state solution becomes less and less optimal. Udaikumar[19] reported where light intermission, and duration are non-limiting to achieve the high growth rate of CO_2, fertilization has a profound effect on increasing growth rate and relating the process of senescence. Short term CO_2 enrichment was shown to increase photosynthetics rate by providing more substrates for photosynthesis and decreased photorespiration. Carbon dioxide is a potent inhibitor of ethylene action and thus reduces the rate of senescence and improves quality of flowers and foliage plants.

2.4. Cooling and Ventilation

Natural ventilation is the direct result of pressure differences created and maintained by wind or temperature gradients. It requires less energy and equipment as well as is the cheapest method of cooling a greenhouse. It depends heavily on evapotranspirational cooling provided by the crop[20]. Teitel et al.[21] investigated the air flow patterns and air temperature distributions in a naturally ventilated greenhouse with vertical roof openings using computational fluid dynamics technique. The results showed a significant effect of the wind direction on the flow patterns both inside and at the roof openings. Wind direction significantly affected the ventilation rate, airflow and crop temperature distributions. Measured ventilation rates are in reasonable agreement with estimated ventilation rates predicted by a model. However, air velocities at the greenhouse openings differed from the measured values.

Evaporative cooling is the most effective cooling method for controlling the temperature and humidity inside a greenhouse. However, its suitability is restricted to the respective region and climate as humid tropics seldom suited for its application due to high humidity levels[20]. By and large, greenhouse users consider fan and pad systems or fan plus foggers as cooling systems[22].

Though various fan and pad systems are applied in greenhouses in many parts of the world today, the fan and fog system is also becoming popular among growers. There are well-defined standards on the greenhouse cooling systems based on fan and pad, and fan and fogging systems[23]. Parallel to the fan and pad systems is the cooling tower application for greenhouse air conditioning[24]. Large or leafy plants such as cucumber can themselves also serve as air coolers due to evapotranspiration from the leaf surfaces, provided that the plants are healthy and have proper irrigation. This type of cooling can be more effective if combined with shade and can decrease fan or ventilator capacity needs, eliminating the pad requirement[25,26].

An experiment was conducted by Kurklu and Bilgin[27] to cool a 15 m^2 ground area plastic-tunnel-type greenhouse by the use of a rock bed. Thermo-physical properties of the rock bed and air are presented in Table 2.1. Two rock-bed canals were dug in the soil, filled with the rocks and insulated; the top surface was covered by a soil layer of thickness enough for the root development depth of the plants. Air was pushed through the rock bed by a centrifugal fan with an 1100 m^3/h flow rate. Energy stored in the rock bed during the day was dumped outside the greenhouse at night using the cooler outside air. The results of the measurements showed that the rock-bed system maintained air temperature 14 °C lower at maximum in the experimental greenhouse than the control one. The temperature difference seemed to increase with increasing solar radiation and outside air temperature. Relative humidity during the day remained at about 40 per cent in the experimental greenhouse and was always higher than that in the control one. The coefficient of performance (COP) of the rock-bed system was higher than 3 in general, and it was observed that this value increased with decreasing rock-bed temperature. The average solar collection efficiency was 38 per cent. The rock-bed system seems to have a significant potential for cooling applications in greenhouses.

Table 2.1: Thermo-physical properties of the rock bed and air

Equivalent diameter	30 mm
Density	1430 kg/m^3 (measured)
Porosity	49% (measured)
Specific heat	0.8 kJ/ kg K (assumed)
Thermal conductivity	2.9 W/m K
Air mass flow rate	0.366 kg/s
Specific heat of air	1012 J/kg K

2.5 Type of Greenhouses

Greenhouses are classified on the basis of different shapes, which also determine their cost, climate control and use in terms of crop production. Commonly used structural designs are briefly described below:

2.5.1 Lean-to Greenhouse

The lean-to greenhouse shares a wall with a building usually the home. It could be garage and relies on the building structure to provide required support to the greenhouse roof (Fig. 2.2). This type of greenhouse provides good insulation to the building. The sun exposed wall is covered by a greenhouse canopy which reduces the heat gain. Such greenhouse is limited to total width of 7 to 12 feet. However, it can be extended as long as attached wall of building. Sometime moist air entered into living areas in hot summer, may create unhealthy environment in living space. Therefore, proper ventilation is desirable, otherwise, dirt and insect may enter into the house.

Fig. 2.2: Lean-to Greenhouse

2.5.2 Even Span Greenhouse

An even-span greenhouse is a structure that is constructed where the roof's pitch is of both equal length and angle. It is one of the most popular greenhouse design because of its relative ease of construction. The optimized design provides sufficient growing space and is normally less costly compared to other greenhouse shapes or forms. A common even-span greenhouse structure that uses arching pipes for the framework is called a hoophouse (Fig. 2.3). Such greenhouse design has a better shape than a lean-to type greenhouse for air circulation to maintain and distribute uniform temperature during the winter heating season.

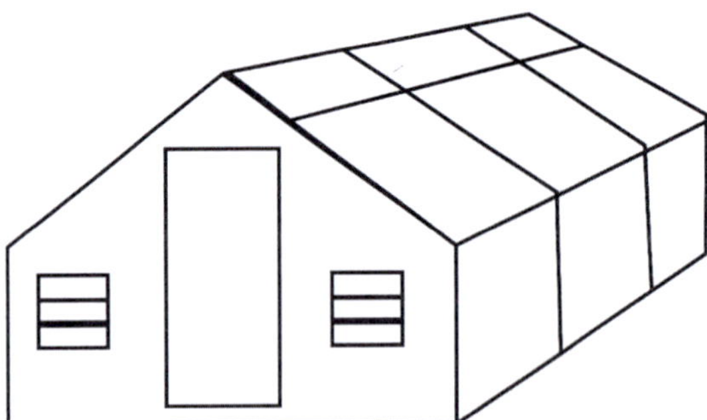

Fig. 2.3: Even Span Greenhouse

2.5.3 Uneven Span Greenhouse

An uneven-span greenhouse is a structure with one roof slope longer than the other. In uneven-span greenhouses, the roof is not equal to width or pitch and usually constructed and adapted in hill terrain. The side that faces the south is transparent and act as a solar collector while opposite is an opaque to conserve the energy (Fig. 2.4). An uneven-span greenhouse is costly compared to a hoop house or an even-span greenhouse because it requires more support on the roof since it is slanted and long.

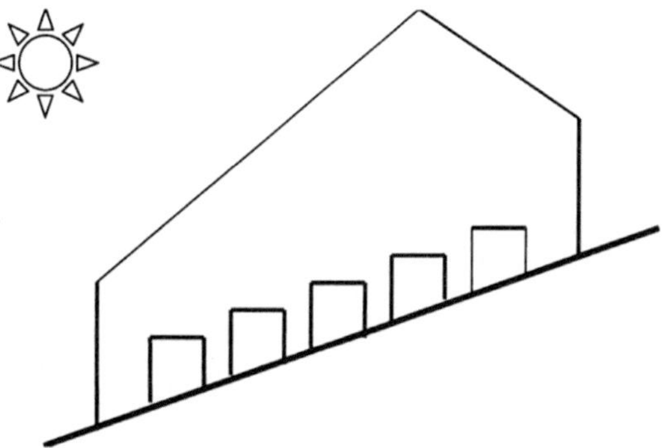

Fig. 2.4: Uneven Span Greenhouse

2.5.4 Ridge-and-Furrow Greenhouse

Ridge-and-furrow greenhouse structures consist of a number of greenhouses connected along the length of the house. The shared interior walls reduce energy costs and allow for large interior spaces. Such greenhouses are best oriented to north and south directions to reduce permanent shadows on the crops, which are created by the gutters (Fig. 2.5). Large interior working space reduces labour and automation cost and improves greenhouse management.

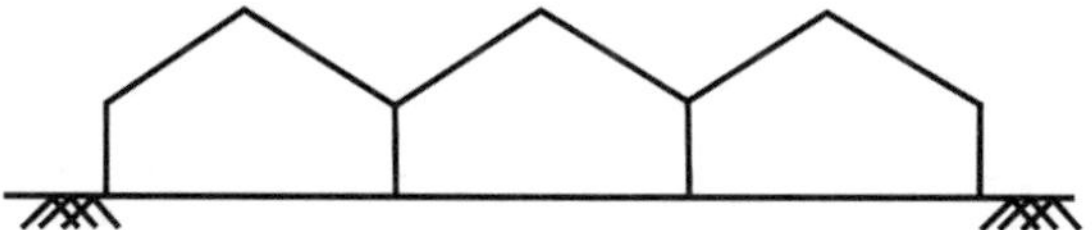

Fig. 2.5: Ridge-and-Furrow Greenhouse

2.5.5 Sawtooth Greenhouse

Sawtooth Greenhouse structures combine optimal ventilation with special strength for withstanding different loads. The roof ventilation alone provides 25 – 30 per cent of the total ventilation of the covered area, in addition to the side ventilation. The shape of the arches allows excellent light transmission.

When open, the sawtooth vent allows a continuous airflow taking place, which reduce the inside temperature or can be closed to optimise the climate control of the growing area (Fig. 2.6).

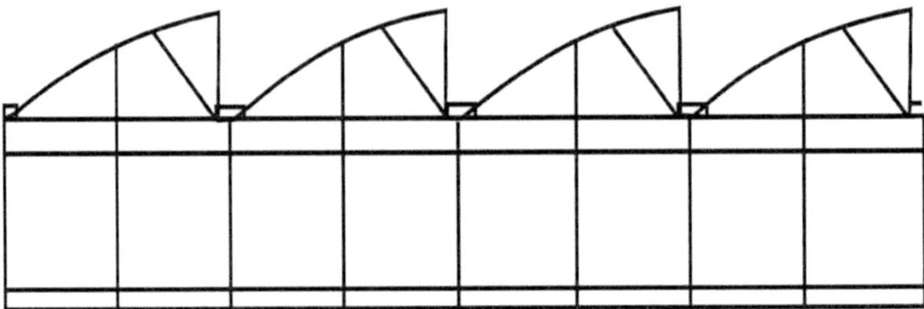

Fig. 2.6: Sawtooth Greenhouse

2.5.6 Quonset Greenhouse

A quonset greenhouse looks like an upside down half circle or oval shape. The frame is made of laminated wood or bent pipe called truss and is covered with polyethylene plastic. The advantage of this house is the ease of construction and covering. Ventilation is by exhaust fans at the ends of the houses (Fig. 2.7)

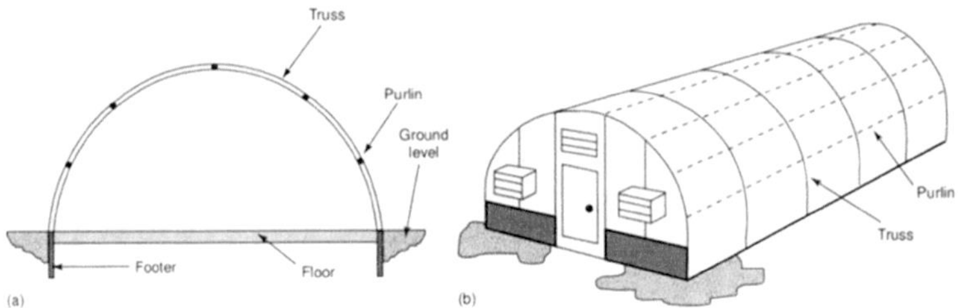

Fig. 2.7: Quonset Greenhouse

2.5.7 Shade Greenhouses

Shade greenhouse is covered by shaded nets which are perforated plastic material used to cut down the solar radiation and prevent scorching or wilting of leaves caused by marked temperature increases within the leaf tissue from strong sunlight. These nets are available in different shading intensities ranging from 25 per cent to 75 per cent. Leafy vegetable and ornamental greens are recommended to be grown under shade net whose growth rates are significantly enhanced to unshaded plants when sunlight is strong.

2.5.8 Net Houses

Shade greenhouse and net greenhouse are often synonymously used but more correctly a net house is enclosed with perforated screen primarily to act as a barrier for the entry of insect and pests. Insect proof nylon nets are available in different intensities of perforations, ranging from 25 to 60 mesh. Net of 40 mesh are effective means to control entry of most flying insect and save the crop from diseases.

2.6 Covering materials of Greenhouses

As mentioned earlier, the purpose of a greenhouse covering is to allow sunlight to pass through it so that the energy is retained inside. Glass was the main covering material in the early greenhouses. With the introduction of plastic materials, there are now several alternatives available for greenhouse covering. A brief description of covering material is given below:

2.6.1 Glass

A clean, transparent glass provides the maximum light transmittance (90 per cent). However, being heavier in weight, it requires elaborate structure for adequate support. It is brittle and can break with minimum shock or vibration resulting in high maintenance costs.

2.6.2 Acrylic

This material has long service life, good light transmittance (80 per cent), moderate impact resistance, but prone to scratches. It has a high coefficient of expansion and contraction. Being inflammable and costly, it is not a preferred material.

2.6.3 Polycarbonate

It is available in single or double wall sheet of different thickness. A new polycarbonate sheet has good light transmittance of about 78 per cent, but reduces with age. It has excellent impact resistance and low inflammable. High cost limits its use on large scale.

2.6.4 Fiberglass reinforced plastic panels (FRP)

The plastic panels consist of polyester resins, stabilized glass fiber, etc. It has a transmittance limit of about 80 per cent and has high impact resistance with a service life ranged from 6-12 years. Good quality FRP materials for greenhouse are not quite assured.

2.6.5 Polyethylene

A clear polyethylene sheet has about 88 per cent light transmittance. Its higher strength and low cost have made it the most popular replacement to glass. An ultra-violate (UV) stabilized plastic sheet can have a service life of 3 years.

These sheets are generally available in 7 and 9 meter widths with 200 micron (0.2 mm) thickness.

2.7 Greenhouse ventilation

Ventilation of an enclosed space is the process of replacing the air inside the space with the ambient or conditioned air. In case of open circuit ventilation, it is the ambient air replacing the enclosed air mass. In the case of closed circuit ventilation, the air from the enclosure is drawn, conditioned and returned to the enclosure. A hybrid ventilation system, working as an open circuit system for a part of the time and as a closed circuit for the rest of the time, is the also used, especially for plant growth chambers.

The need for greenhouse ventilation arises for the control of one or the following parameters:

1. Temperature
2. Humidity
3. Carbon dioxide concentration
4. Air circulation around plants

It may be noted that ventilation is not the only method of controlling the above four parameters rather it is an important means to achieve the above mentioned parameters.

2.7.1 Ventilation requirements

Ventilation requirements would vary depending on whether it is for the control of temperature, moisture, CO2, any other gas in the greenhouse air, or for maintaining air circulation around plants. Ventilation requirement may be determined either by rules of thumb or by analytical procedures. The well accepted rule of thumb for it is as follows:

1. Provide an airflow rate of 0.03 – 0.04 cu.m/sq.m. or 6 – 8 cfm/sq ft of floor space.
2. Provide one volume air exchange per minute of ventilation during summer season.
3. Airflow rate in the greenhouse should be such that air velocity around the plants is maintained in the range of 0.5 to 1.0 m/s.

The above mentioned norms are in general, convenient and help in determining capacity of the ventilation system. However, for more precise estimation, analytical methods may be used.

2.8 Methods of Ventilation

There are two methods of ventilation, namely: natural and forced. Natural ventilation refers to provide sufficient open area in the greenhouse structure so that ambient air by itself enters into the greenhouse after displacing an equal amount of greenhouse air. In the forced ventilation system, auxiliary power is used to move air through the greenhouse. A brief description of these methods is given below:

2.8.1 Natural Ventilation

Many older greenhouses depend on natural ventilation for air movement. However, with increased concern about the high cost of energy required to operate greenhouses today, increased emphasis is again being placed on natural ventilation systems for greenhouses. In greenhouses employing natural ventilation systems, sidewall vents and ridge vents continue for the full length of the building can be opened as far as it is desired to allow air to move through the house. To be ventilated satisfactorily, the house must have both sidewall and ridge vents. If a house has only side vents, then it can only be ventilated during periods of wind movement outside. Using ridge vents and side vents permits the greenhouse to be vented by both wind pressure and thermal gradients. Thermal gradients generally are created within the greenhouse by solar energy heating the materials inside, which in turn heat the air. As air is heated, it becomes lighter and rises through the ridge vents, with the makeup air coming from outside through the sidewall vents. If sidewall and ridge vents are properly sized, quite satisfactory ventilation rates can be achieved with some degree of temperature control. A natural ventilation system will not be as dependable or satisfactory as a mechanical ventilation system in terms of providing continuous, uniform greenhouse ventilation. Some newly-designed greenhouses with natural ventilation systems are capable of achieving a high degree of environmental modification for increased plant production. However, for those plant species that require air temperatures lower than outside air temperature, evaporative cooling must be used. Mechanical ventilation is an integral part of any evaporative cooling system.

2.8.2 Effect of insect-proof screens on Ventilation

The use of an insect proof screen in greenhouse reduces the insect attacks and their migration on the crop resulting reduces the crop damage[28]. It also reduces the pesticide demand. However, it reduced effect of natural ventilation hence increment in the greenhouse inside temperature[29]. It increases the pressure drop on the opening, resulting in poor ventilation, and it is directly concerned with screen porosity. The screen porosity (ε) can be calculated from the geometric dimensions of the screen using in given formula:

$$\varepsilon = \frac{(\delta - d)(m - d)}{m\delta}$$

where

δ and m are the distance between the centres of two adjacent weft and warp threads, respectively, and d is the distance of the threads.

An antithrips screen can reduce ventilation by approximately 60–70 per cent while an antiaphid screen can reduce it by 40 per cent[30]. In the present context, insect screen application has been improved using optical insect preventions with insect screen, and removing screen when insect risk is low. Ventilation reduction can be improved by increasing the ventilation surface and by increasing the screen area. Screens with a smaller thread diameter also improve the ventilation rate as they are more porous.

2.9 Greenhouse Cooling System

The need to cool a greenhouse arises whenever the greenhouse air temperature crosses the upper limit of the crop tolerance. Failure to bring down the temperature effectively may result in either partial or total crop failure within a very short period of time. Design of an appropriate cooling system for greenhouse operation in most parts of India is essential because the cooling season may last upto to eight months annually.

2.9.1 Fan and Pad Evaporative Cooling Systems

Fan and pad systems consist of exhaust fans at one end of the greenhouse and a pump circulating water through and over a porous pad (Fig 2.8) installed at the opposite end of the greenhouse.

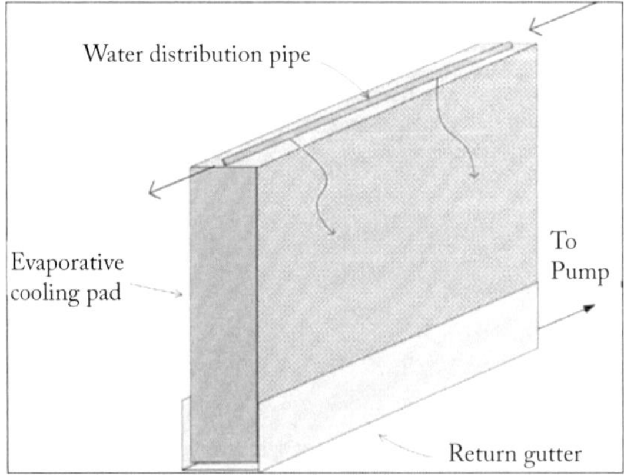

Fig. 2.8: Evaporative Cooling Pad

Water flows along the distribution pipe and drains down into the pad material. The sump should be large enough to hold all run-off when the pump is turned off.

If all vents and doors are closed when the fans operate, air is pulled through the wetted pads and water evaporates. Removing energy from the air lowers the temperature of the air being introduced into the greenhouse.

The air will be at its lowest temperature immediately after passing through the pads. As the air moves across the house to the fans, the air picks up heat from solar radiation, plants, and soil, and the temperature of the air gradually increases. The resulting temperature increase as air moves down the greenhouse produces a temperature gradient across the length of the greenhouse, with the pad side being coolest and the fan side warmest.

2.10 Greenhouse Control Systems

Once a greenhouse structure is built various techniques, devices etc. must be added in order to control the environment. Control systems include those for lighting, heating, cooling, relative humidity and carbon dioxide enrichment.

2.10.1 Light

Maximum light transmission, of the appropriate quantity and quality (photosynthetically active radiation, 400-700 nm), through the greenhouse structure to the plants is crucial for optimum photosynthesis, growth and yield. Excess light can be reduced by using shade paint or using external shade cloth with varying degree of mesh size.

2.10.2 Heating

Heating device is desirable to maintain an optimum temperature for plant growth during cold weather. The basic heating system consists of a fuel burner, heat exchanger, distribution system and controls. Heat delivery to the crop is done by convection and radiation. Insulating material (cloth or film curtains) can be positioned above the crop or near the roof to retain heat near the crop. The insulating material used during the night can be the same material used for shading during the day.

2.11 Types of Heat Loss from a Greenhouse

Conduction: Heat transfer either through an object or between objects in contact. Conduction depends on area, path length, temperature differential and physical properties of the object(s).

Convection: Heat transfer by the movement of warm gas or liquid to a colder location. Convection depends on temperature differential. Example: Movement of warm air near the plants upward towards the roof.

Radiation: Heat transfer between separated objects. Radiation occurs from all objects and depends on the areas, temperatures and surface characteristics of the objects involved.

2.12 Relative Humidity

High or low relative humidity can be detrimental to plant growth. When RH is too high, transpiration (the movement of water from inside the leaf to the outside) is reduced along with movement of mineral nutrients, and the pollen can clump on the stigma causing cat facing or the pollen may not be released from the anthers at all. When RH is too low, transpiration may be increased significantly resulting in plant wilt and decreasing pollination. Relative humidity can be increased by running the cooling pads or by fogging whereas it can be running the heaters or simply venting.

2.13 Thermal Modeling of Greenhouse

Thermal model of a greenhouse for growing vegetable crops during off season has been developed. The shading nets were provided inside the greenhouse, which divide the greenhouse in two parts, hence two temperature layers are considered inside it. One is above the shading net, and second one is below the shading net, where actual cultivation operation is being performed. Thermal flux generated in typical greenhouse is illustrated in Fig. 2.9.

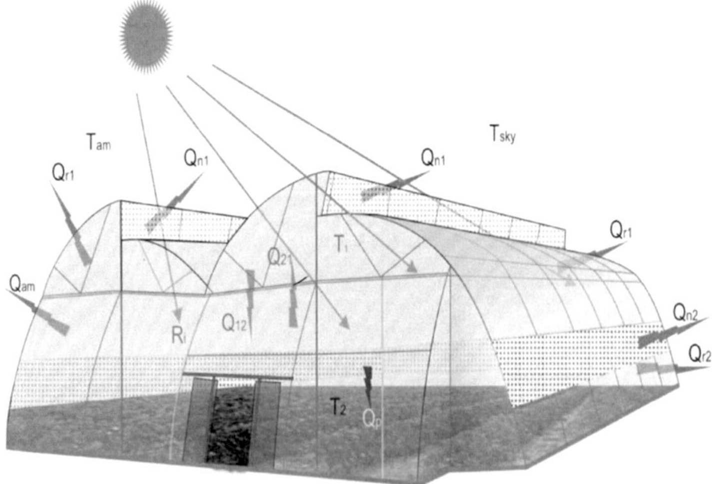

Fig. 2.9: Schematic diagram of Greenhouse Thermal Model

In order to write an energy balance equation for each component of the greenhouse, the following assumptions have been made.

1. The greenhouse air is well mixed at all times so that no temperature or moisture gradient exists in the air.
2. The temperatures of the internal air, crops and cover materials are in a steady state.
3. Crop is planted on sowing bed, the width of bed is kept 75 cm and the distance between two beds is kept 30 cm.
4. The heat transfer from greenhouse air to floor is neglected.

Chen et al.[31] developed a model which reveals the effect of temperature in the greenhouse with shading nets. They cover greenhouse with one external shading net, and shading is also provided inside the greenhouse. The present model is inspired from their study and applied it for composite climatic condition of Udaipur in Indian state of Rajasthan.

2.13.1 Thermal Flux for Upper Layer

Solar energy is main source of energy input. The net input thermal flux at upper layer depends on transmittance of greenhouse cover is as follows:

$$Qin_1 = \tau o\, A_1\, I_s \qquad (2.1)$$

Energy flow to ambient atmosphere by heat transfer

$$Qam_1 = A_1\, h_{gc}\, (T_1 - T_{am}) \qquad (2.2)$$

Shading networks as thermal screen and there is scope of heat thermal interaction between upper and lower layer. This can be presented as follows:

$$Q_{12} = A_{shd}\, U_{shd}\, (T_1 - T_2) \qquad (2.3)$$

Energy exchange by natural ventilation is given by

$$Q_{n1} = AR_1\, \rho\, C_p\, (T_1 - T_{am}) \qquad (2.4)$$

AR_1 is natural air exchange rate as adapted from Roy et al.[32]

$$AR_1 = \frac{A_{op1}}{2} C_d \left[2g \frac{\Delta T}{T_{am}} \frac{H_1}{4} \right]^{0.5}$$

$$AR_1 = 0.5\, A_{op1}\, C_d \left[g \frac{(T_{av} - T_{am}) H_1}{2T_{am}} \right]^{0.5}$$

$$T_{av} = \frac{(T_1 V_1 + T_2 V_2)}{V_1 + V_2}$$

Radiative heat transfer at upper layer is given by

$$Q_{r1} = \varepsilon_{gc} F_{1s} \sigma A_{shd} (T_1^{\,4} - T^4_{sky}) \quad (2.5)$$

The correlation with sky temperature (T_s) and ambient temperature (T_{am}) is adapted from Diffie and Becman[33]

$$T_{sky} = 0.0552 (T_{am})^{1.5}$$

Thus the energy balance equation for the upper layer is

$$\tau_o A_1 I_s = A_1 h_{gc} (T_1 - T_{am}) + A_{shd} h_{shd} (T_1 - T_2) + AR_1 \rho C_p (T_1 - T_{am}) + \varepsilon_{gc} F_{1s} \sigma A_{shd} (T_1^{\,4} - T^4_{sky}) \quad (2.6)$$

2.13.2 Heat Transfer Model of the Lower Layer

Input flux at lower layer depends on the transmittance of both greenhouse cover and shading net, hence the net input solar flux reaching at lower layer is given by;

$$Qin_2 = \tau_0 \tau_{in} I_s A_2 \quad (2.7)$$

Energy flow to ambient atmosphere by heat transfer is

$$Qam_2 = A_2 h_{gc} (T_2 - T_{am}) \quad (2.8)$$

The energy exchange between upper and lower layer is

$$Q_{21} = A_{shd} U_{shd} (T_2 - T_1) \quad (2.9)$$

Energy exchange by natural ventilation from lower layer is

$$Q_{n2} = AR_2 \rho C_p (T_2 - T_{am}) \quad (2.10)$$

$$AR_2 = 0.5 A_{op2} C_d \left[g \frac{(T_{av} - T_{am}) H_2}{2T_{am}} \right]^{0.5}$$

Energy exchange by natural ventilation at lower layer by upper layer vent is

$$Q_{n12} = AR_1 \rho C_p (T_1 - T_2) \quad (2.11)$$

Radiative heat transfer at lower layer

$$Q_{n12} = AR_1 \rho C_p (T_1 - T_2) \quad (2.12)$$

Absorption of thermal energy by shading net is

$$Q_{r2} = \varepsilon_{shd} F_{2s} \sigma A_f (T_2^{\,4} - T^4_{sky}) \quad (2.13)$$

Heat transfer due to crop transpiration is given by

$$Q_{ab} = \alpha A_{shd} (1 - \tau_{in}) \tau_0 I_s \quad (2.14)$$

The transpiration model of tomatoes is adopted from the HORTITRANS model[34]

$$E_t = \frac{aI_s}{\lambda} + \frac{h_t}{\lambda\gamma}\left(P_{ws} - \frac{P_{vp}}{1000}\right)$$

$a = 0.154 \ln (1 + 1.1\ LAI^{1.3})$

$$h_t = 1.65\ LAI\left[1 - 0.56\exp\left(\frac{-R_i}{13.0}\right)\right]$$

$R_i = \tau_0\ \tau_{in}\ I_s$

$\gamma = 0.066$ k Pa K^{-1}

$\lambda = 2260$ kJ kg^{-1}

Saturation vapour pressure (P_{ws}) is calculated by Weiss[35] expression

$P_{ws} = 0.61078 \exp (17.2694\ T_2)/(T_2 + 237.3)$

The temperature dependent partial vapor pressure can be evaluated by the following expression[36]:

$$P_{vp} = \exp\left(25.317 - \frac{5144}{T_2}\right) \times 10^{-3}$$

Thus, the energy balance equation for lower layer is given by

$$\tau_0\ \tau_{in}\ I_s\ A_2 = A_2\ h_{gc}\ (T_2 - T_{am}) + A_{shd}\ h_{shd}\ (T_1 - T_2) + AR_2\ \rho\ C_p\ (T_2 - T_{am}) + AR_1\ \rho\ C_p\ (T_1 - T_2) + \varepsilon_{shd}\ F_{2s}\ \sigma\ A_f\ (T_2^4 - T_{sky}^4) + \alpha\ A_{shd}\ (1 - \tau_{in})\ \tau_0\ I_s + \lambda\ E_t\ A_f\ P_f \quad (2.15)$$

2.13.3 Solution Procedure

Matlab software tool can be used for computing the unknown parameters T_1 & T_2 of equations 2.6 and 2.15. The flow chart of solution is illustrated in Fig. 2.10. The input parameters are given in Table 2.2.

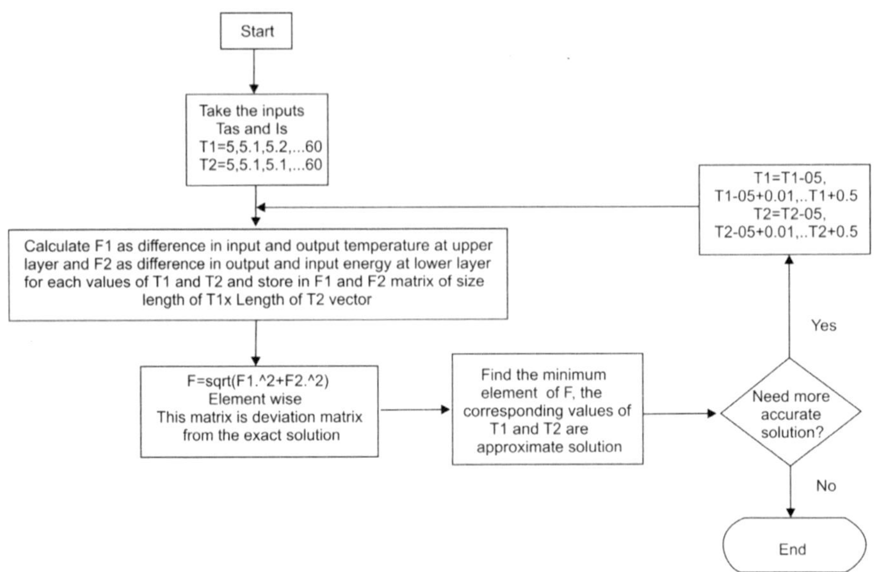

Fig. 2.10: Flow chart of Mathematical Model of Naturally Ventilation Greenhouse

Table 2.2: Input Parameters for Greenhouse

C_d	0.35
C_p	1005 J kg^{-1} °C^{-1}
F_{1s}	0.71
F_{2s}	0.79
LAI	2
α	0.04
ρ	1.2 kg m^{-3}
σ	5.67 x 10^{-8} W m^{-2} K^{-4}
t_0	0.75
t_1	0.5
λ	2.257 J kg^{-1}
γ	66 Pa K^{-1}
ε_{gc}	0.9
ε_{shd}	0.6

2.14 Energy and Exergy Analysis of Greenhouse

2.14.1 Energy Analysis

The solar energy available inside the greenhouse is responsible for increasing the greenhouse air temperature. An increment in the temperature improves the microclimatic conditions of greenhouse air. The energy input is mainly depended on available solar radiation, greenhouse area and cover transmissivity of greenhouse cover.

The energy input to the greenhouse can be written as:

$Q_{in} = \tau_0\ \tau_{in}\ I_s\ A_2$

Energy output depends on the temperature difference between greenhouse air and ambient conditions. It can be expressed as:

$Q_{Out} = \dot{m}_a\ C_a\ (T_2 - T_{am})$

Energy efficiency

$$\eta = \frac{Q_{out}}{Q_{in}} \qquad (2.16)$$

2.14.2 Exergy Analysis of Greenhouse

The term exergy is defined as the maximum amount of useful work that can be obtained from a system[37,38]. The rational efficiency based on the concept of exergy is a true measure of the performance of a solar thermal system. The energy efficiency is the quantitative assessment of energy whereas exergy is the qualitative assessment of energy. The exergy input at lower layer of greenhouse can be written as:

$$Ex_{in} = \tau_0\ \tau_{in}\ A_{gc}\ I_s\left[1 - \frac{4}{3}\left(\frac{T_{am}}{T_s}\right) + \frac{1}{3}\left(\frac{T_{am}}{T_s}\right)^4\right]$$

Where T_s=6000 K

Exergy output

$$Ex_{out} = \dot{m}_a C_a (T_2 - T_{am})\left[1 - \frac{T_{am}}{T_2}\right]$$

Exergy efficiency

$$\eta_{II} = \frac{Ex_{out}}{Ex_{in}} \qquad (2.17)$$

2.14.3 Thermal Exergy Transfer

The exergy transfer at lower layer is calculated by estimating the exergy losses in different heat transfer modes in greenhouse. The scope of exergy losses in lower layer is in ventilation at both upper and lower layer, heat transfer due to long wave radiation; heat transferred to be ambient, crop transpiration and some unaccountable losses inside the greenhouse is also considered. The unaccountable exergy losses may be due to heat transfer to the ground through conduction and plant canopy and bed etc. The exergy losses in various heat transfer routes at lower layer of greenhouse are estimated as follows:

Exergy loss during ventilation at lower layer

$$Ex_{Ln1} = Q_{n1}\left[1 - \frac{T_{am}}{T_2}\right] \tag{2.18}$$

Exergy loss during ventilation at upper layer

$$Ex_{Ln21} = Q_{n21}\left[1 - \frac{T_{am}}{T_2}\right] \tag{2.19}$$

Exergy loss due heat transfer to ambient

$$Ex_{Lam} = Q_{am}\left[1 - \frac{T_{am}}{T_2}\right] \tag{2.20}$$

Exergy loss during heat transfer due to long wave radiation

$$Ex_{Lr2} = Q_{r2}\left[1 - \frac{T_{am}}{T_2}\right] \tag{2.21}$$

Unaccountable thermal exergy losses

$$Ex_{Lua} = Q_{ua}\left[1 - \frac{T_{am}}{T_2}\right] \tag{2.22}$$

Exergy transfer due to heat transfer from lower layer to upper layer

$$Ex_{T21} = Q_{21}\left[1 - \frac{T_2}{T_1}\right] \tag{2.23}$$

Exergy transfer due heat absorbed by shading net

$$Ex_{Tab} = Q_{ab}\left[1 - \frac{T_{am}}{T_{shading}}\right] \tag{2.24}$$

Where,

$$T_{shading} = \frac{T_1 + T_2}{2}$$

Exergy transfer due to heat absorbed by plant during transpiration

$$Ex_{Tp} = Q_{p2}\left[1 - \frac{T_{am}}{T_p}\right] \tag{2.25}$$

Where,

$T_p = T_2$

Total exergy losses can be written as:

$$\sum Ex_L = Ex_{Ln1} + Ex_{Ln21} + Ex_{Lam} + Ex_{Lr2} + Ex_{Lua}$$

Total exergy transfer can be written as:

$$\sum Ex_T = Ex_{T21} + Ex_{Tab} + Ex_{Tp}$$

The exergy balance equation can be written as:

$$Ex_{in} - (Ex_{out} + \sum Ex_L + \sum Ex_T) = Ex_{Dest} \quad (2.26)$$

2.15 Statistical Analysis of Developed Model

In order to assess the consistencies between predicted and measured air temperature, a statistical analysis has been carried out. The standard error (*SE*), Root Mean Square Error (*RSME*) and coefficient of correlation (*r*) parameters used in the study are defined as Holman[39]:

Standard error (*SE*)

$$SE = \frac{\sigma}{\sqrt{n}} \quad (2.29)$$

Where

$$\sigma = \left[\frac{\sum_{i=1}^{n}\left(x_{exp} - x_{pre}\right)^2}{n-1}\right]^{1/2}$$

Root Mean Square Error (*RMSE*)

$$RMSE = \frac{1}{n}\sqrt{\sum_{i=1}^{n}\left[x_{i,\,exp} - x_{i,\,pre}\right]^2} \quad (2.30)$$

A relationship for the correlation coefficient (*r*) which may be preferable is as follows:

$$r = \frac{n\sum x_{exp}x_{pre} - \left(\Sigma x_{exp}\right)\left(\Sigma x_{pre}\right)}{\left[n\sum x_{exp}^2 - \left(\sum x_{exp}\right)^2\right]^{1/2}\left[n\sum x_{pre}^2 - \left(\sum x_{pre}\right)^2\right]^{1/2}} \quad (2.31)$$

Where Xexp and Xpre are experimentally recorded and theoretical greenhouse air temperature, n is the number of observation.

2.16 Thermal Performance of Greenhouse

The developed model can predict greenhouse air temperature. Temperature inside the greenhouse for both upper and lower layer was recorded at regular intervals of time from 9:00 hours to 17:00 hours. It was found that temperature at upper layer is always higher than corresponding to lower layer. The shading net used inside the greenhouse which reduces the transmission level of solar radiation to lower layer that is why the temperature at lower layer is lower than upper layer. The difference between ambient and lower layer temperature varied from 2-5 °C for the months January-April, 2012 during these months and ambient temperature was varying from 12 – 35 °C. Theoretically calculated greenhouse air temperature at lower layer is 1-3 °C higher than experimental values. Similar trend was found for the month July- December. But greenhouse air temperature during months May - June is 8- 11°C higher than ambient temperature as shown in Fig. 2.11 and Fig. 2.12 respectively, such difference can be reduced by using misting system, but it required frequent cooling. Scarcity of water and irregular electricity supply does not allow for using greenhouse during these months in Rajasthan state.

Statistical analysis between theoretical and experimentally recorded greenhouse air temperature shows good correlation. The values of the correlation coefficients (*r*) for lower layer are greater than 0.97 for all the months; it demonstrates that, the proposed model can provide good fitness with experimental values.

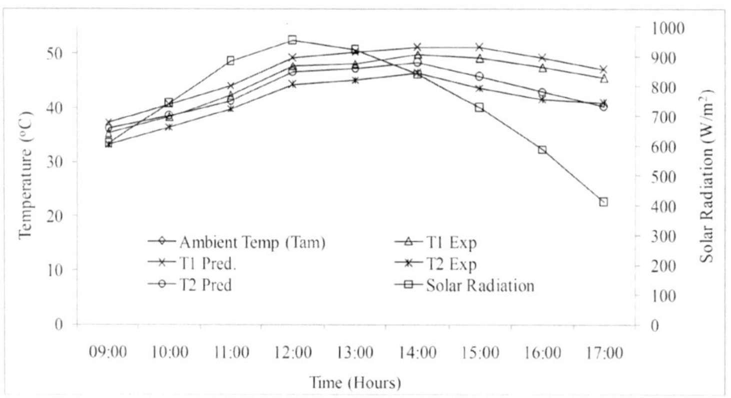

Fig. 2.11: Experimental and Predicted Greenhouse Air Temperature for the Month May, 2012

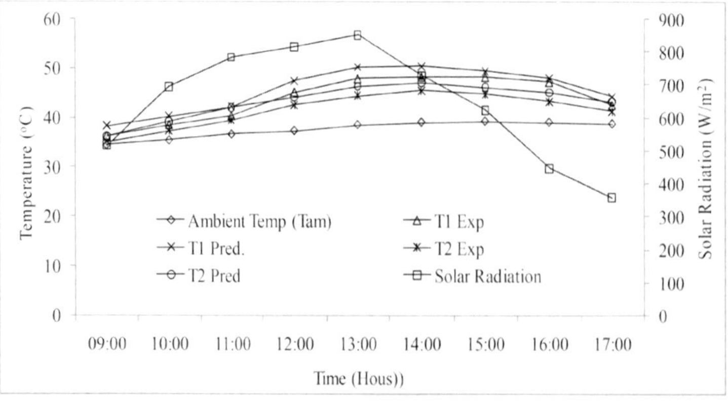

Fig. 2.12: Experimental and Predicted Greenhouse Air Temperature for the Month June, 2012

2.17 Energy and Exergy Analysis

Exergy is defined as the maximum useful amount of work that can be produced by a system or a flow of matter or energy as it comes to equilibrium with a reference environment[37]. Exergy is a true measurement of the quality or grade of energy, and it can be destroyed in the thermal system. In the present study, both energy and exergy analysis was carried out for lower layer of greenhouse with the equations (2.16 - 2.17). The predicted energy and exergy efficiencies for all the months were higher than that of experimentally calculated. Month January and December show higher energy and exergy efficiency at 17:00 hours in spite of low solar radiation, because of maintaining high inside air temperature with the help of shading net, which acts as a thermal screen. Hence, the naturally ventilated type greenhouse can provide favourable air temperature during winter season.

In protected cultivation practices, such as structure is constructed for providing desirable air temperature, relative humidity and solar radiation. Even though with high solar radiation, greenhouse air temperature can be reduced by misting, more air exchange, etc. the energy loss from each mode of heat transfer are calculated to identify the accountable energy losses. The average experimental and theoretical energy efficiency varied from 1.71 - 3.63 per cent respectively and 2.70 – 6.50 per cent respectively, and corresponding average exergy efficiency varied from 0.021- 0.046 per cent and 0.048 – 0.124 per cent respectively.

Experimental energy efficiency during the winter month of December, 2011 and January, 2012 was 3.325 per cent and 3.634 per cent and exergy efficiency was varying from 0.038 per cent to 0.046 per cent as presented in Table 2.3. It is a fact that both energy and exergy output is the function of temperature whereas the input is the function of collector area and solar insolation. Though, solar greenhouse is having more solar collector area but less temperature rises as compared to other solar thermal devices. This is the main reason for low energy and exergy efficiency. In spite of having low efficiencies such a system is capable to provide favourable microclimatic condition for plant propagation.

Table 2.3: Energy and Exergy Efficiency of Naturally Ventilated Greenhouse

Months	Energy Efficiency (%)		Exergy Efficiency (%)	
	Exp.	Pred.	Exp.	Pred.
July, 2011	2.941	4.983	0.039	0.108
Aug.,2011	2.887	4.401	0.040	0.083
Sept., 2011	2.662	4.057	0.035	0.075
Oct. ,2011	2.544	3.889	0.034	0.071
Nov., 2011	2.869	4.572	0.036	0.074
Dec., 2011	3.325	6.185	0.038	0.107
Jan. 2012	3.634	6.502	0.046	0.124
Feb. 2012	2.810	4.392	0.035	0.083
March, 2012	1.716	2.701	0.021	0.048
April, 2012	1.916	2.900	0.024	0.049
May, 2012	2.140	3.098	0.035	0.066
June, 2012	2.586	3.847	0.040	0.079

2.18 Exergy Transfer

The heat loss is calculated by using equation 2.18 – 2.25. Equation 2.26 is used to estimate total exergy transfer in greenhouse. Fig 2.13 reveals the theoretical exergy flow during the month March, 2012. The exergy transfer 4919.04 kJ/day was found in crop transpiration, and it is about 1.126 per cent of exergy input. Ventilation at upper layer lifted the heat from lower layer of greenhouse during this process about 1304.13 kJ/day exergy is being lost. Exergy loss at lower layer ventilation is about, 94.74 kJ/day. Total theoretical exergy transfer during the month March was calculated as 428527.31 kJ/day, which is about 98.11 per cent of exergy input.

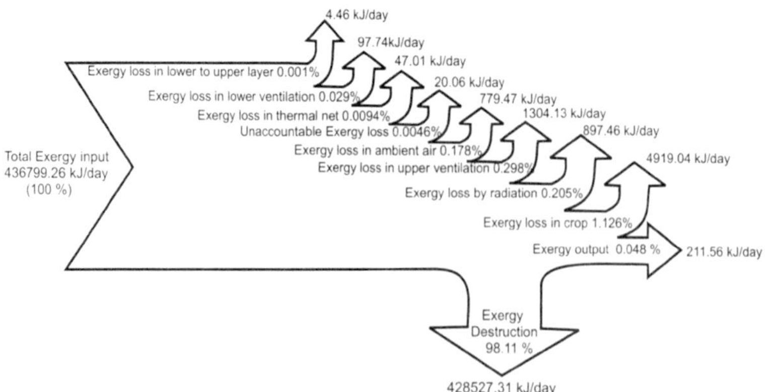

Fig. 2.13: Theoretical Exergy Flow During the Month March, 2012

Experimental exergy transfer during the March, 2012 in different heat transfer routes are demonstrated in Fig.2.14. The maximum energy is utilized by plant during transpiration; during this process, 3666.40 kJ/day of exergy is being lost. Unaccountable thermal exergy losses were calculated about 93.76 kJ/day, and it is 0.021 per cent of exergy input. Total exergy transfer is estimated 431884.67 kJ/day, which is about 98.87 per cent of exergy input.

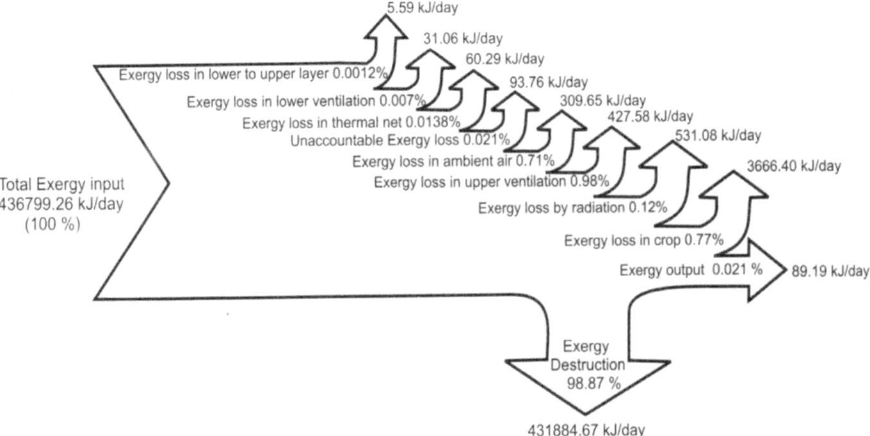

Fig. 2.14: Experimental Exergy flow during the month March, 2012

Nomenclature

A_1	surface are of upper layer, m^2
A_2	wall surface are of lower layer, m^2
A_f	floor area, m^2
A_{op1}	opening area of upper layer, m^2
A_{op2}	opening area of lower layer, m^2
A_{shd}	shading net area, m^2
AR_1	Natural air exchange rate in upper layer, m^3s^{-1}
AR_2	Natural air exchange rate in lower layer, m^3s^{-1}
C_d	discharge coefficient
C_p	air specific heat, J $kg^{-1}\,{}^{o}C^{-1}$
E_t	transpiration rate, kg $m^{-2}s^{-1}$
F_{1s}	shape factors for the sky as seen from the upper layer
F_{2s}	shape factors for the sky as seen from the lower layer
g	gravitational acceleration, ms^{-1}
H_1	opening height of upper layer, m
H_2	opening height of lower layer, m
h_{gc}	convective heat transfer coefficient between upper layer and ambient air, W m^2 K^{-1}
h_{shd}	convective heat transfer coefficient between upper layer and lower layer, W m^2 K^{-1}

I_s	solar radiation, Wm^{-2}
LAI	leaf area index
P_{ws}	air saturated vapour pressure inside air, kPa
P_{vp}	Vapour pressure inside the air, kPa
Rh	relative humidity,%
R_i	incoming solar radiation at lower layer of the greenhouse, W m^{-2}
T_1	upper layer temperature, °C
T_2	lower layer temperature, °C
T_{am}	ambient temperature
T_{ave}	average temperature of greenhouse, °C
T_{sky}	sky temperature, K
V_1	volume of the upper layer, m^3
V_2	volume of the lower layer, m^3
α	absorptance of shading nets
ρ	air density, kg m^{-3}
σ	Stephan-Boltzman constant, W m^{-2} K^{-4}
ε_{gc}	thermal emittance of cover materials
ε_{shd}	thermal emittance of shading net materials
t_0	transmittance of greenhouse cover
t_1	transmittance of shading nets
λ	latent heat of vaporization, kJ kg^{-1}
γ	thermodynamic psychrometric constant, kPa K^{-1}

References

1. Panwar NL. Performance evaluation of greenhouse for cultivation of medicinal plant in composite climate of Udaipur. Unpublished ME thesis. Udaipur, Rajasthan, India: Maharana Pratap University of Agriculture and Technology; 2001.
2. Cooper PI, Fuller RJ. A transient model of the interaction between crop, environment and greenhouse structure for predicting crop yield and energy consumption. J Agric Eng Res 1983;28:401–17.
3. Mahrer Y, Avissar R. A numerical simulation of the greenhouse microclimate. Math Comput Simulat 1984;XXVI:218–28.
4. Feuilloley P, Issanchou G. Greenhouse covering materials measurement and modelling of thermal properties using the hot box method, and condensation effects. J Agric Eng Res 1996;65:129–42.
5. Jain D, Tiwari GN. Modeling and optimal design of evaporative cooling system in controlled environment greenhouse. Energy Convers Manage 2002;43:2235–50.
6. Kendirli B. Structural analysis of greenhouses: a case study in Turkey. Build Environ 2006;41:864–71.
7. Panwar NL, Kothari S, Rathore NS. Protected cultivation of medicinal plant in composite climate of Indian State Rajasthan. American-Eurasian J Agric Environ Sci 2009;5(5):633–7.
8. Bot GPA. Developments in indoor sustainable plant production with emphasis on energy saving. Comput Electron Agric 2001;30:151–65.
9. Straten G. Acceptance of optimal operation and control methods for greenhouse cultivation. Annu Rev Control 1999;23:83–90.

10. Castan eda-Miranda R, Ventura-Ramos E, Peniche-Vera RDR, Herrera-Ruiz G. Fuzzy greenhouse climate control system based on a field programmable gate array. Biosyst Eng 2006;94:165–77.
11. Luo W, De Zwart HF, Dai J, Wang X, Stanghellini C, Bu C. Simulation of greenhouse management in the subtropics—Part I: Model validation and scenario study for the winter season. Biosyst Eng 2005;90:307–18.
12. Brown HT, Escombe F. The influence of varying amounts of carbon dioxide in the air on the photosynthetic process of leaves and on the mode of growth of plants. Proc R Soc Lond 1902;70:397–413.
13. Owen O, Small T, Williams PH. Carbon dioxide in relation to glasshouse crops. Part III. The effect of enriched atmospheres on tomatoes and cucumber. Ann Appl Biol 1926;13:560–76.
14. Bolas BD, Melville R. The effect on the tomato plant of carbon dioxide produced by combustion. Ann Appl Biol 1935; 22:1–15.
15. Cummings MB, Jones CH. The aerial fertilization of plants with carbon dioxide. Vermont Agric Exp Stn Bull 1916; 211:1–56.
16. Mortensen LM. Review: CO2 enrichment in greenhouse. Crop response. Sci Hortic 1987; 33:1–25.
17. Shaw RJ, Roger MN. Interaction between elevated carbon dioxide levels and greenhouse temperature on growth of roses. Hort Rev 1964;135(3486):23–4.
18. Ioslovich I, Segnier I, Gutman PO, Borshchevsky M. Sub-optimal CO2 enrichment of greenhouse. J Agric Eng Res 1995; 60:117–36.
19. Udaikumar L. Carbon dioxide enrichment in greenhouse influence on productivity and quality of flowering plants. In: International proceeding on protected cultivation. 1997.
20. Kumar KS, Tiwari KN, Jha MK. Design and technology for greenhouse cooling in tropical and subtropical regions: a review. Energy Build 2009; 41:1269–75.
21. Teitel M, Liran O, Tanny J, Barak M. Wind driven ventilation of a mono-span greenhouse with a rose crop and continuous screened side vents and its effect on flow patterns and microclimate. Biosyst Eng 2008;101(1):111–22.
22. Arbel A, Yekutieli O, Barak M. Performance of a fog system for cooling greenhouses. J Agric Eng Res 1999; 72:129–36.
23. ASABE Standards. Heating, ventilating and cooling greenhouses, EP406.4 JAN2003 (R2008):819.
24. DeJong T, VanDeBraak NJ, Bot GPA. A wet plate heat exchanger for conditioning closed greenhouses. J Agric Eng Res 1993;56(1):25–37.
25. Yang X, Short TH, Fox RD, Bauerle WL. The microclimate and transpiration of a cucumber crop. Trans ASAE 1989;32(6):2143–50.
26. Fynn RP, Al-Shooshan A, Short TH, McMahon RW. Evapotranspiration measurement and modeling for a potted chrysanthemum crop. Trans ASAE 1993;36(6):1907–13.
27. Kurklu A, Bilgin S. Cooling of a polyethylene tunnel type greenhouse by means of a rock bed. Renew Energy 2004; 29:2077–86.
28. Teitel M, Barak M, Berlinger MJ, Lebiush-Mordechai S. Insect proof screens: their effect on roof ventilationand insect penetration. Acta Horticulturae 1999; 507: 29–37
29. Soni P, Salokhe VM, Tantau HJ. Effect of screen meshsize on vertical temperature distribution in naturallyventilated tropical greenhouses. Biosystems Engineering 2005; 92(4): 469–482.
30. Pérez Parra J, Baeza E, Montero JI, Bailey BJ. Natural ventilation of Parral greenhouses. Biosys. Eng. 2004; 87(3): 355–366.
31. Chen C, Shen T, Weng Y. (2011) Simple model to study the effect of temperature on the greenhouse with shading nets. African Journal of Biotechnology 2011; 10(25): 5001-5014.

32. Roy JC, Boulard Y, Kittas C, Wang S. (2002) Convective and ventilation transfers in greenhouses, Part 1: the greenhouse considered as a perfectly stirred tank. Biosyst. Eng. 2002; 83: 1-20.
33. Duffie JA, Beckman WA. Solar engineering of thermal processes. New York: John Wiley & Sons. 1991.
34. Jolliet O. Hortitrans, a model for predicting and optimizing humidity and transpiration in greenhouse in greenhouse. J Agric Eng Res 1994; 57: 23–37.
35. Wiess A. Algorithms for the calculation of moist air properties on a hand calculator. Amer. Soc. Ag. Eng. 1997; 23: 1133-1136.
36. Fernandez J, Chargoy N. Multistage indirectly heated solar still. Solar Energy 1990; 44: 215-223.
37. Dincer, I. (2000) Thermodynamics, exergy and environmental impact. Energy Sour 2000; 22: 723-732.
38. Panwar NL, Kaushik SC, Kothari S. (2012) A review on energy and exergy analysis of solar dying systems. Renew Sust Energ Rev 2012; 16: 2812-2819.
39. Holman JP. Experimental Methods for Engineers. Tata McGraw-Hill, India 2010.

3

Vertical Farming

3.1 Introduction

The world population will grow by another 2 billion people by 2025 and feeding them will be a big challenge. Rapid industrial development and fast urbanization, the arable lands are continuously shrinking. Many scientists and researchers reported that the Earth had lost a third of its arable lands over the last 40 years[1]. In the current scenario, the most challenging task for agricultural sciences is to ensure for continuous and enough supply of food to growing human civilization[2]. Presently, all available arable land on the earth are using extensive pesticides and fertilizers in various ways that are not sustainable as well as harmful to the environment.

There is an urgent need to think about sustainable farming options as arable land for farming is continuously reducing to fulfil the food material demand of growing population. Vertical farming can produce more food from fewer land and water resources. Though it takes out many of the risks inherent in outdoor crop production. With the control of light exposure, indoor temperature, relative humidity and watering levels, growers can grow food very efficiently. Vertical farming has the potential to produce food with less energy, less water, less waste and in less space than traditional methods. It also negates the need for harmful chemical fertilizers and pesticides. Developing and installing vertical farm in those areas where people need good quality healthy food, tackle the shortage of nutritive food material at a local level. Vertical farming can be defined as it is the practice of growing crops in vertically stacked layers where controlled-environment agriculture practices usually incorporated with the aims to optimize plant growth. The soilless farming techniques such as hydroponics, aquaponics, and aeroponics also incorporated in vertical farming as shown in Fig. 3.1. Vertical farming is a very old idea, and indigenous peoples used vertically layered growing techniques like the rice terraces of East Asia. The term vertical farming was coined by American geologist Gilbert Ellis Bailey in 1915. In 1999, Dickson Despommier, a professor at New York's Columbia University, popularized the modern idea of vertical farming, building upon the idea together with his students[3].

Despommier and his students came up with a design of a 30 story crop powerhouse that could feed 50,000 people[4]. Although the design has not yet been built, it successfully popularized the idea of vertical farming. Jeffrey Landau, director of business development at Agritecture Consulting estimates the global value of the vertical farming market will rise to about $6.4 bn by 2023, from $403m in 2013, with almost half that attributed to growth in the US[5].

Fig. 3.1: Vertical Farming System

3.2 Types of Vertical Farms

Advancement in the field of science and technology is the driving force of advance research in the field of Indian Agriculture, which enables Indian farmers to feed. The growing population. Indian government emphasis on double the income of farmers by 2022. This can be achieved with increasing productivity, minimising losses, minimal input and high output, and effective use of arable land. Vertical farming is option where non arable land, abandoned building, etc. can be used for growing crops through different techniques. There are different shape and size of vertical farming, it may be wall mounted, and multi-level building. All types of vertical farming approach using any one of three soilless media to provide desired nutrients to plant for proper growth – hydroponics, aeroponics, or aquaponics.

3.2.1. Hydroponics

Hydroponics normally refers to the technology where plants are growing without soil. In such growing system plants, roots are submerged in liquid solutions containing micronutrients, which is frequently monitored and circulated to ensure that the correct chemical composition is maintained as shown in Fig. 3.2. In addition to this, inert (chemically inactive) media such as pebbles, sand, and sawdust are used to provide support to the roots[6].

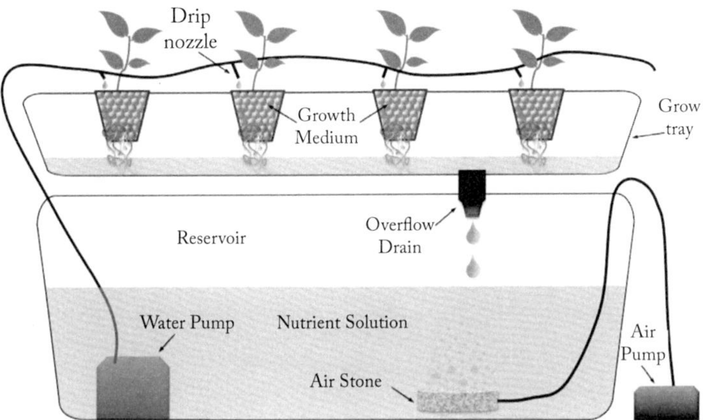

Fig. 3.2: Hydroponic system

3.3.2. Aeroponics

Aeroponics refers to the techniques of growing plants in an air or mist environment without the use of soil or an aggregate medium as shown in Fig. 3.3. It is efficient and sustainable soil-less plant growing technique for vertical farming, using up to 90 per cent less water than even the most efficient hydroponic systems[7]. Aeroponic is highly recommended in vertical farming because it saves energy as gravity automatically drains away excess liquid[8]

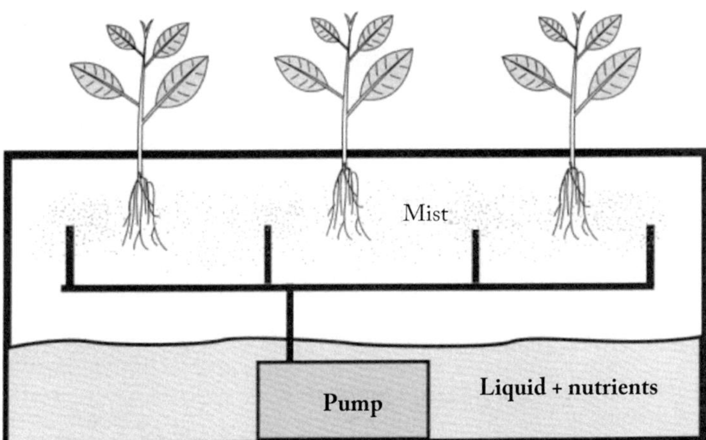

Fig. 3.3: Aeroponic System

3.2.3. Aquaponics

The term aquaponics is formed by combining two words: aquaculture, which refers to fish farming, and hydroponics—the technique of growing plants without soil. It is the combination of growing plants and fish in the same ecosystem as shown in Fig. 3.4. Nutrient-rich wastewater from the fish tanks is filtered by a solid removal unit and then led to a bio-filter, where toxic ammonia is converted to nutritious nitrate. While absorbing nutrients, the plants then purify the wastewater, which is recycled back to the fish tanks. Moreover, the plants consume carbon dioxide produced by the fish, and water in the fish tanks obtains heat and helps the greenhouse maintain temperature at night to save energy[9]. Although, aquaponics is used in smaller-scale vertical farming systems, most commercial vertical farm systems focus on producing only a few fast-growing vegetable crops and don't include an aquaponics component[7].

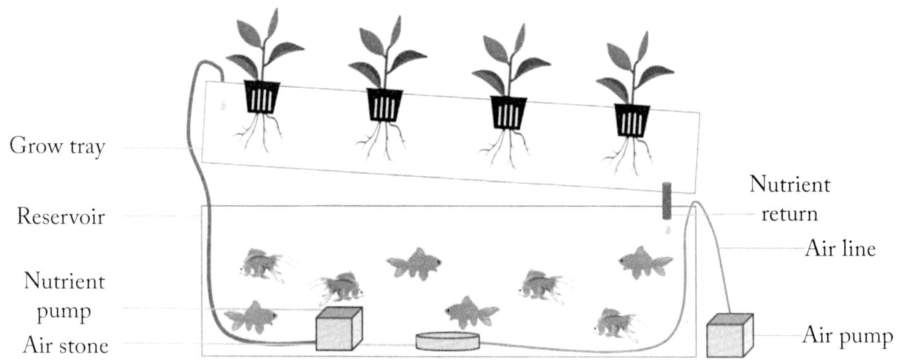

Fig. 3.4: Aquaponic System

3.3 Factors Affecting Success of Vertical Farm

The vertical farming approach is promoted with the aims to increase productivity with limited space, produce a clean and green source of food along with biosecurity, and minimal use of transportation and fossil fuels. As discuss earlier that the vertical farming model is essentially an indoor farm based on a high-rise multi-level building design. The success of vertical farming highly depends on following factors[10]:

3.3.1 Effective Space Utilization and Productivity

The aim of vertical farming approach is to maximizes space use not only by using one layer of the growing space, but utilizing the entire volume of space, from the floor to up as illustrated in Fig. 3.5. It provides an effective engineering solution to increase productivity per unit area through spreading plant cultivation into the vertical dimension, it enhances land use efficiency for crop production[11]. Despommier suggested that a 30 story building of 100 m height with a basal area of 2.02 ha would be able to produce a crop yield equivalent to 971.2 ha of conventional horizontal farming[12]. Thus, maximizing space use, growers must maximize growing surface area within a volume. While assessing productivity of vertical farming both growing and access space should be considered separately. Estimate access space and space use efficiency with a ratio that compares production space to floor space.

Fig. 3.5: Effective Space Utilization in Vertical Farming

3.3.2 Labour Costs

Enormous power and labour cost involvement are one of the main challenges for the vertical farming industry. To reduce the labour cost, one can think about the fully automated vertical farming system but initial investment is extremely high. The labour and operating cost can be reduced by improving efficiency and using indigenous automation facility. Information technology based plant growth solution can be a viable solution to improve efficiency and reduce labour cost. On the other hand, to reduce the labour cost, growers should maximise both the ease with which workers can access crops, and the number of workers that are able to access a layout simultaneously. This will certainly reduce the waiting time and improve the efficiency. The maintenance process is to be streamlined. In the vertical farming system, workers can easily identify the plants problem and easily interact with issues to resolve them, which affect product quality and profit.

3.3.3 Profitability

Vertical farming is being seen as the way forward to produce higher volumes of better-quality crops all year round in limited space, bring food production closer to customers, and into urban areas. It is normally assumed that vertical farming occupies more crop volume hence there will be more profit. Factually, more volume will not always be profitable. Filling more volume requires more monitoring and controlling the micro climatic condition for proper plant growth. It increases the operational cost when compared to traditional farming.

Ricardo Hernandez, horticulture professor at North Carolina State University reported that vertical farm is economically viable for growing transplants or starter plants. The starter plants are a high value product, and they can be grown under very high density in vertical farms, even higher than they can be grown in a greenhouse. These transplant can be sold to greenhouse and field growers who will produce the end products[13]. Surekha reported that in vertical farming approach 3 to 4 times higher yield can be achieved. Moreover, with a crop cycle of 3-4 times a year, the profits are substantial as compared to traditional farming[14]. Vertical farming uses a lot of electricity to perform different unit operation. Solar photovoltaic based electricity generation system significantly cut the electricity bill and contributes increment in net profit.

3.3.4 Environment and Plant Health

There are several factors that influence the growth rate and health of crops. These factors most often include:

- Temperature
- Humidity
- Carbon dioxide
- Light
- Nutrient concentration
- Nutrient pH

In order to effectively manage growing conditions in vertical farming, controlled environment agriculture conditions are to be maintained and monitored. Controlled environment agriculture can be defined as a method of cultivating plants in an enclosed environment, using technology to ensure optimal growing conditions. The controlled environment agriculture cultivation process can be done in virtually any form of contained area, and it is being looked towards as a viable alternative for modern food production.

3.4 Feasibility of Vertical Farming in India

India is one of the largest producers of vegetables, fruits and many other agricultural commodities and has one of the largest arable lands in the world. Vertical farming is the best thing and future of Indian agriculture business as India's population is increasing with every second, and land is limited. Many scientists, researchers, and academician reported that tillage is harmful for land, and under no tillage farming pesticides and herbicides are extensively used to protect the plants which is harmful for human beings. Vertical farming is just opposite to that. It uses a small amount of land, and also one can perform organic farming.

Vertical farming is limited in India at present to high value crops only. Indian Institute of Agriculture Research (IIAR) is working on techniques and adaptations to introduce in the market. It is also in vogue for production of disease free crop. Most common and successful vertical farming example is mushroom cultivation. Vertical farming is a great investment as far as Indian agricultural economy is concerned. There are many big challenges for the Indian agricultural community in present scenario, which are killing the value of agricultural produces. Due to improper storage facilities, and lack of expert advice on crop management, agricultural produce through vertical faming is bound to fail. The scope of vertical farming is however, increasing gradually in India. Good technical and financial support is desirable for establishing vertical farming units.

3.5 Converting Greenhouse into Vertical Farming System

Vertical Farming is one of the agricultural systems where the yield received per unit area is high. Effective energy utilization and sustainable production in vertical farming system, engineering, architecture, technology and experiences are needed to be used all together. The production received in a traditional horizontal agricultural system in the area of 4000 - 30000 m^2 can be achieved in the area of 1000 m^2 vertical farming system. Utilization of renewable energy resources can be incorporated while designing the vertical farming systems to reduce environmental pollution and fossil fuel consumption[15].

An effective architectural design of the vertical farming system to supply food material of 15,000 people was proposed by Banerjee and Adenaeuer[16]. The proposed size of vertical farming is 0.93 ha, 37 floors, 25 floors is exclusively recommended for crop production, 3 floors for aquaculture. Further, 3 floors reserved for maintaining and controlling essential micro climatic parameter for plant growth, 2 floors in the basement used for waste management. Furthermore, one floor is kept for cleaning of growth trays, sowing and germination, one floor is allocated for packing and processing the plants and fish, and one floor is reserved for sales and delivery at the basement as shown in Fig. 3.6. Total height of proposed building is about 167.5 m, length and width equal to 44 m. the aspect ratio is kept 3.81. In the present context number of fancy and attractive building designs with sustainability perspectives are available. To achieve sustainability in vertical farming system designers must consider the environment, the weather, orientation, materials, sunlight, health, etc.

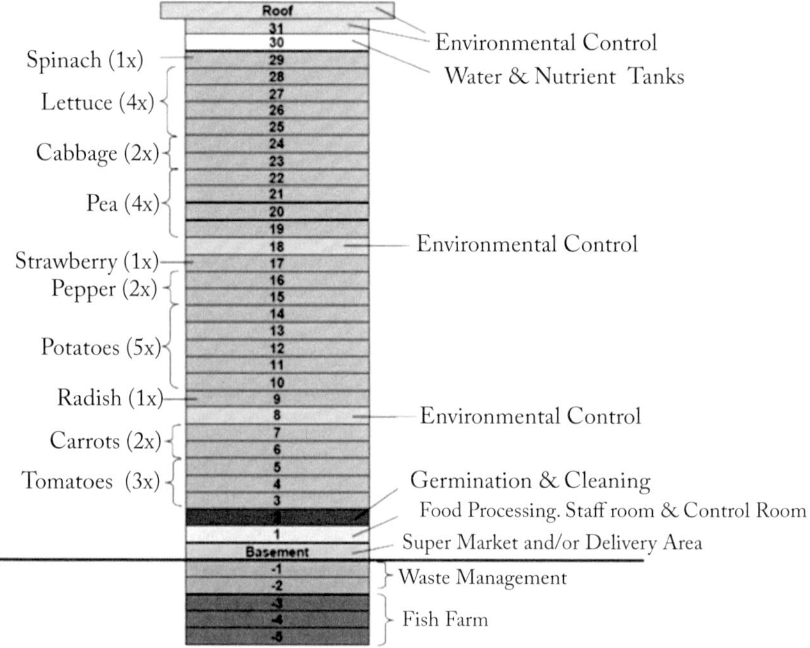

Fig. 3.6: Vertical Farming System[16]

3.6 Construction Material for Vertical Farming

The proper selection of materials for construction of vertical farming system is becoming a focus for designers because of the indirect effect on the environment. Growers are preferring sustainable building, which reduces environmental impact and providing a healthy living space and reduce energy bill. Presently, Double-Skin Facades (DFS) material are extensively used in skyscraper and office building, it improves natural lighting without energy losses. The DFS system comprises of three components:

(i) Exterior Wall

(ii) Ventilated cavity

(iii) Interior Wall.

The exterior wall provides protection against the weather. It is usually a single layer of heat-strengthened safety glass or laminated safety glass. The cavity is an undivided or divided buffer space. It ranges from 5 to 50 inches in width. The cavity insulates the building against wind, sound, and temperature. It pulls in outside air and uses it for the controlled ventilation of interior spaces. The interior wall of DFS acts as thermal insulation[17].

3.7 Lighting

Lighting is a very important component for the vertical farming system. Plants growth purely depends on quality of available light as it used by plants for photosynthesis. Photosynthesis is the process by which plants convert light energy into chemical energy. Various types of artificial light system have been designed to provide light in greenhouses and in vertical farming system. Photo synthetically Active Radiation (PAR) is desirable for photosynthesis and plant growth. The photosynthetic organisms are able to use spectral range of solar radiation from 400 to 700 nm in the photosynthesis process as shown in Fig. 3.7. PAR values range from 0 to 3,000 millimoles per square meter. At night, PAR is zero. During mid-day in the summer, PAR often reaches 2,000 to 3,000 mill moles per square meter.

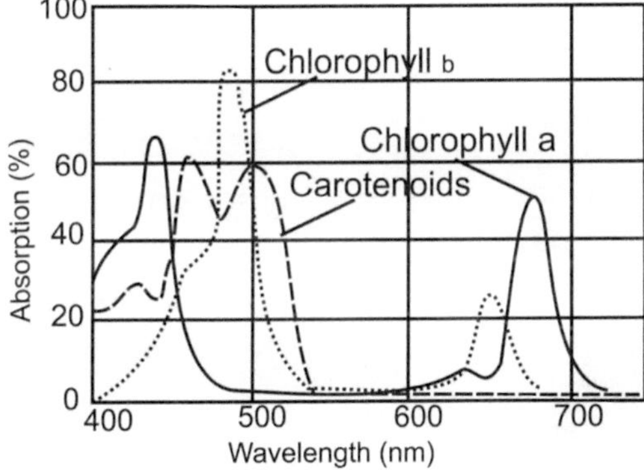

Fig. 3.7: Photosynthetically Active Radiation

3.7.1 Natural Lighting

Natural lighting, also known as day lighting, is a technique that efficiently brings natural light into a home using exterior glazing, thereby reducing artificial lighting requirements and saving energy. Natural lighting has been proven to increase health and comfort levels for building occupants and reduce energy cost[18]. The process of day lighting includes controlling how much natural light (both diffuse and direct) enters a building[19]. Essentially, day lighting is a way to provide energy services without using fuels, just the direct use of primary energy flows. In skyfarming, daylight will decrease as the stacking density within the building cubature increases, and it will be a primary light source at the outer skirt of skyfarm. The adaptability of skyfarm in different location and different climatic conditions highly depends on modular construction schemes, using different architectural cubature concepts and different façade modules incorporating light guiding concepts[20]. Through lifting the main food production field high up in the air, the vegetation gains more exposure toward the natural sunlight and fresh air as shown in first design of Plantagon plantscaper in Linköping, Sweden (Fig. 3.8). The building angle and shape to be followed while designing skyfarm for maximum gain of solar energy as shown in preview of dynamic construction program of Plantagon in Sweden (Fig. 3.9) the ground level becomes more freed up with nicely shaded open spaces which could be enjoyed by the public.

Fig. 3.8: First Design of Plantagon plantscaper in Linköping, Sweden

Fig. 3.9: Swedish Skyfarm Using Maximum Sunlight

3.7.2 Light Emitting Diode (LED)

With increase in the adaptation of vertical farming system, large number of advancements have been made in Light-Emitting Diode (LED) technology in recent years. For last many decades, indoor farming was highly relied on artificial light which were very expensive to operate, and as a result, were used to grow only the most profitable crops. However, today LEDs are so affordable and so efficient that they are dramatically redefining the economics of indoor farming. Productivity of indoor farming highly depends on interior micro climatic conditions, quality of interior light, and minimum the interaction with the exterior climate. Restriction in interaction with exterior climate can also encourage the efficient use of energy, water and carbon dioxide[22].

Plants can utilize light basically in two different bands of frequencies. High intensity bulbs are full spectrum, so they cover all the frequencies. Light Emitting Diodes (LED) lights can only cover a single frequency. LED grow lights typically have two different sets of bulbs so it can produce light in both bands as one band promotes growth, the other promotes flowering. The particular band of the light spectrum is required for indoor farming is in the range of 400 – 700 nm and this band is known as Photo synthetically Active Radiation. The plant primarily used blue light in the range of 400 -490 nm during vegetative growth phase while orange-red light in the range of 580 – 700 nm used during flowering and fruiting phase. Typical arrangement of LED lighting system is vertical farming system is illustrated in Fig. 3.10. LED systems are the perfect solution for vertical farming system because, they are highly energy efficient, about 40 to 70 per cent more energy efficient than high-pressure sodium (HPS) lights or metal halide (MH) lamps. The lifespan of a typical LED bulb has about 50,000 hours of usable life, which translates to nearly six years of continuous use[23].

3.8 Water Requirement

Vertical farming is done under controlled environment conditions. Water is essentially needed to grow plants, cool the ambient air with micro-sprays, washing and cleaning, and other purposes. In vertical farming system usually, hydroponic production system is preferred, which requires only 10 per cent to 16 per cent of the water to produce the same volume of vegetables as conventional irrigation farming methods do in the open field conditions. In traditional irrigation system, water gets lost due to evaporation. The evaporation is a natural procedure but there exists a bigger problem which is the water that goes out of farms as run-off which is useless for drinking since it is polluted by fertilizers, salts, pesticides[24] etc. A full-fledged hydroponic system requires between 5 to 7 litres of water/m^2/day. This excludes water for evaporative cooling, washing, cleaning and other purposes. Fig. 3.11 shows the typical irrigation system in a vertical farming system.

Fig. 3.10: LED Grow Lighting in Vertical Farming

Fig. 3.11: Irrigation System for Vertical Farming

According to the Food and Agriculture Organization (FAO) of the United Nations, the vertical farming system consumes 75 per cent less raw material than traditional farming and just 60 watt of power daily to grow 150 kg of vegetables in a month. To obtain this desired quantity, vertical farming needs just 6 square meter space while traditional farming requires at least 72 square meter land area. Also, the requirement of water would be just 12 litres to produce 1 kg of vegetables due to recycling as against 300-400 litre under traditional farming[25].

3.9 Sustainability

Sustainable food system can be defined as "*a food system that delivers food security and nutrition for all in such a way that the economic, social and environmental bases to generate food security and nutrition for future generations are not compromised*"[26]. Vertical farming system is a solution for creating a complete sustainable food system, on economic and social bases. Basically, sustainability depends on the effective use of locally available resources such as water and land. Vertical farming system uses these resources more effectively than greenhouses and conventional agricultural methods. Higher crop yield can achieve in vertical farming system, if lettuce grow on an agricultural field of 1x1 meter, 3.9 kg of lettuce can be produced every year. Vertical farms can even yield twenty times more lettuce than agricultural fields[27]. The vertical farms lower overall CO_2 output by 67-92 per cent compared with greenhouses[21]. The difference is much larger with field crops[28]. Local production also skips several steps in the supply chain. This benefits the crops' freshness since their travel time is reduced by approximately one week. Local production also improves the traceability of crops and overall reduces its agricultural waste[29].

3.10 Renewable Energy in Vertical Farming

The energy efficiency in vertical farming is higher than greenhouse. However, vertical farming consumes about 30 – 176 kWh per kg more energy than greenhouse[21]. Vertical farming moving towards maximum utilization of renewable energy sources, can reduce the carbon footprint.

Renewable energy and farming are a winning combination. Long term income source to farmers can be provided by harvesting energy from wind, solar, and biomass. Such energy sources are capable to replace conventional energy sources used on the farm.

Wind energy extensively used to generate electricity and to pump water. Solar energy can be used in number of ways to reduce pollution and increasing self-reliance. Solar energy can be used to generated electricity using solar photovoltaic system to power farm operation such as water pumping, lights, etc., which cut the energy bill. Solar heat collector can be used to dry the farm produces and to heat the space. Further, solar water heater can be used to wash the farm produces and to clean the skyfarm floors. Installation of solar panels above a skyfarm enables a grower to produce both electricity and food at the same time as shown in Fig. 3.12.

Fig. 3.12: Cylindrical Towers Outfitted with Rooftop Solar Panels (Designed By Chris Jacob)

Vertical farming offers a good opportunity to obtain more biomass per unit area compared to traditional horizontal farming. It reduces the land footprint of expanded biomass production. The produced waste biomass can be converted into energy by burning, and heat can be utilized for heating building, drying agricultural produces, and process heat for different unit operations. Biomass can be converted into liquid or gaseous fuel to generate electricity or transportation fuel.

Gordon Graff, a Toronto-based architecture researcher designed a vertical farm and proposed 59 storeys which could produce food material for 50,000 household. Further, Graff estimated energy requirement for lighting propose is about 82 GWh per annum. Graff proposed that half of this energy could be generated by combustion of methane generated from the farm own waste. Building biomass waste biogas digester is viable option to generate biogas.

3.11 Role of Information Technology

The vertical farming system is basically an indoor farm based on a high-rise multi-level factory design. Effective use of rainwater and recycling water, maintain and control micro climatic parameter such as temperature, humidity, and LED lighting and heating. Desired photosynthesis spectrum from LED lighting can be programmed using information technology[30]. The vertical farming system is no stranger to information technology. The past decade has seen dramatic advances in electronic hardware and software, transforming society's abilities to collect, store, process and transmit useful information[31]. There are many tools can be considered under the subject heading, Information Technology (IT) since, by definition, IT is concerned with the acquisition, recording and communication of information[32].

Indian based IT solution provided company TheGreenSAGE™ developed "Smart Automated Grow Environment" IoT solution that enables individuals and commercial growers to conveniently grow fresh, flavorful culinary herbs throughout the year. TheGreenSAGE™ solution provides optimal conditions for efficiently growing culinary herbs & greens so that the users does not require prior knowledge of growing. TheGreenSAGE™ solution provides automatic monitoring and control of nutrients, irrigation, light and other critical parameters required during different stages of optimal plant growth. TheGreenSAGE™ solution is powered by Bitmantis Cloud Platform™ & other indoor/precision agriculture technologies to automate growth functions like Nutrient Management, Irrigation Management, Light Management and Grow Zone management[33]. Information Technology Companies can play a significant role in building such integrated system, which farmers can use remotely to completely control the relevant parameters. IoT framework can play a major role by encompassing and integrating three functional layers namely; sensing layer, delivery layer and control layer[34]. Fig. 3.13 shows few examples of analysis of different layers.

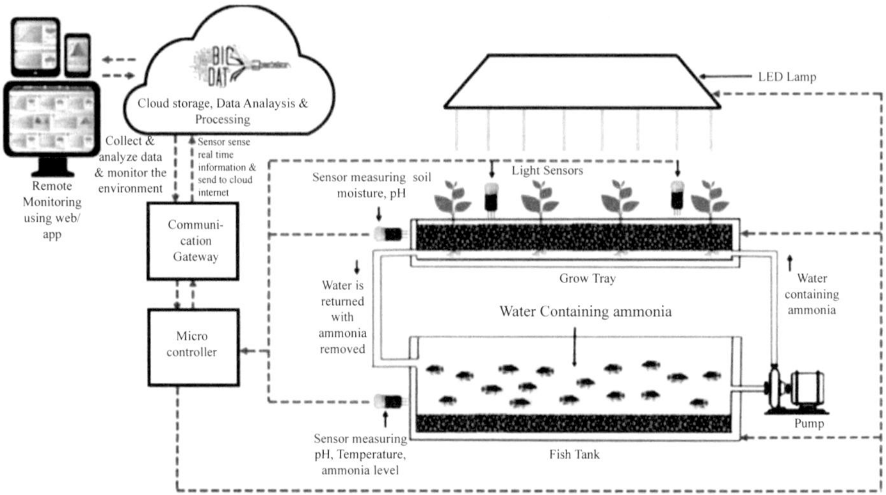

Fig. 3.13: Schematic Diagram of Implementation of IoT in Vertical Farming[34]

3.12 Advantages and Disadvantages of Vertical Farming

Advantages

(a) Year-round crop production is possible in the vertical farming system.

(b) It creates new urban employment opportunities.

(c) It significantly reduces the use of conventional sources of energy.

(d) Vertical farming is capable to provide nutritional food material to metropolitan cities;

(e) There are no weather related crop failures.

(f) Vertical farming system is capable of converting black and gray water into potable water by engineering intervention.

(g) It adds energy back to the grid via methane generation from anaerobic digestion of non-edible parts of plants and animals.

(h) It provides sustainable environments in urban areas.

(i) Abandoned multi-floors urban building can be converted into food production centres.

(j) No use of pesticides.

(k) It allows maximum crop yield.

(l) It reduces transportation costs.

(m) It uses very minimal water.

(n) Vertical farming requires less area to grow crops/plants.

Disadvantages

(a) There will be fewer jobs as people do not need persons for transporting the crops.

(b) There will be lack of pollinators in the crops, wages paid for pollution will be very high as it has to be done manually.

(c) Vertical farming system has complex machinery and automated systems. Therefore, vertical farming requires more energy input compared to field farming.

(d) Highly dependent on technology. If a vertical farm loses power for one day, then it will be a big loss in production.

References

1. Leblanc R. What You Should Know About Vertical Farming. Is It the Future of Agriculture? https://www.thebalancesmb.com/what-you-should-know-about-vertical-farming-4144786 accessed on May 22, 2020.
2. Garg A, Balodi R. Recent trends in agriculture vertical farming and organic farming. Adv Plants Agric Res. 2014;1(4):142–144.
3. Lawson B. Opinion on Vertical farming: from Babylon to New York. https://www.cambridgeconsultants.com/insights/opinion/vertical-farming-babylon-new-york accessed on May22, 2020.
4. Cooper A. Going up? Vertical Farming in High-Rises Raises Hopes. https://psmag.com/environment/farming-in-high-rises-raises-hopes-3705 accessed on May22, 2020.
5. Hotten R. The future of food: Why farming is moving indoors. https://www.bbc.com/news/business-49052317, accessed on May22, 2020.
6. Resh HM. Hydroponic food production: a definitive guidebook for the advanced home gardener and the commercial hydroponic grower. CRC Press; 2012
7. Birkby J. Vertical Farming. ATTRA Sustainable Agriculture Program. https://attra.ncat.org/product/vertical-farming/ Retrieved May 23, 2020.
8. Mytton-Mills H. Reimagining Resources to Build Smart Futures: An Agritech Case Study of Aeroponics. InSmart Futures, Challenges of Urbanisation, and Social Sustainability 2018: 169-191. Springer, Cham.
9. Kledal PR, Thorarinsdottir R. Aquaponics: a commercial niche for sustainable modern aquaculture. In Sustainable Aquaculture 2018: 173-190. Springer, Cham.
10. Hamarikrishi web page: https://hamarikrishi.com/the-4-factors-of-vertical-farm-success/ accessed on May 24, 2020
11. Eigenbrod C, Gruda N. Urban vegetable for food security in cities. A review. Agron. Sustain. Dev. 2014; 35: 483–498.
12. Al-Kodmany K. The vertical farm: A review of developments and implications for the vertical city. Buildings. 2018; 8(2):24
13. Kuack D. Can vertical farms be profitable? Urban Ag News https://urbanagnews.com/blog/exclusives/can-vertical-farms-be-profitable/ accessed on May 25, 2020.

14. Surekha Kadapa-Bose. Soil-less & Vertical Farming. https://www.thebetterindia.com/199411/how-to-vertical-farming-hydroponics-fresh-produce-organic-india/ accessed on May22, 2020.
15. Cıceklı M, Barlas NT. Transformation of today greenhouses into high technology Vertical Farming Systems for metropolitan regions. Journal of Environmental Protection and Ecology. 2014;15(4):1779-85.
16. Banerjee C, Adenaeuer L. Up, up and away! The economics of vertical farming. Journal of Agricultural Studies. 2014; 2(1):40-60.
17. Double-Skin Façade System: Materials, Advantages & Examples. (2017) Retrieved from https://study.com/academy/lesson/double-skin-fa-ade-system-materials-advantages-examples.html.
18. Boyle G. Renewable Energy: Power for a Sustainable Future, 2nd ed. Oxford, UK: Oxford University Press, 2004.
19. Gregg D. Ander F. Daylight: Whole Building Design Guide. (September, 15, 2016). https://www.wbdg.org/resources/daylighting.php accessed on May 26, 2020.
20. Germer J, Sauerborn J, Asch F, de Boer J, Schreiber J, Weber G, Müller J. Skyfarming an ecological innovation to enhance global food security. J. Verbr. Lebensm. 2011; 6:237–251.
21. Graamans L, Baeza E, Van Den Dobbelsteen A, Tsafaras I, Stanghellini, C. Plant factories versus greenhouses: Comparison of resource use efficiency. Agricultural Systems 2018; 160: 31-43.
22. Goto E. Plant production in a closed plant factory with artificial lighting. Acta Hortic. 2012; 956: 37–49.
23. Crumpacker M. 4 Facts You Need to Know about LEDs and Vertical Farming (Nov. 9, 2018) https://medium.com/@MarkCrumpacker/4-facts-you-need-to-know-about-leds-and-vertical-farming-bb1bed18064d accessed on May 27, 2020.
24. Despommier D. The rise of vertical farms, *Sci. Am. 2009;* 301(5): 80–87.
25. Rural Marketing. Vertical farming: Securing food and nutrition for growing population. https://ruralmarketing.in/industry/agriculture/vertical-farming-securing-food-and-nutrition-for-growing-population Accessed on May 27, 2020.
26. High Level Panel of Experts on Food Security and Nutrition (HLPE) (2014). Food Losses and Waste in the Context of Sustainable Food Systems, Report of the HLPE, Rome: HLPE
27. Bayley JE, Yu M, Frediani K. Sustainable food production using high density vertical growing (Verticrop). In XXVIII International Horticultural Congress on Science and Horticulture for People (IHC2010): International Symposium 2010: 95-104.
28. Kozai T, Niu G, Takagaki M. (Eds.). Plant factory: an indoor vertical farming system for efficient quality food production. Academic Press. 2015.
29. Naus T. Is Vertical Farming Really Sustainable?. Eit Food. August 29, 2018 https://www.eitfood.eu/blog/post/is-vertical-farming-really-sustainable accessed on May 27,2020
30. Benke K, Tomkins B. Future food-production systems: vertical farming and controlled-environment agriculture. Sustainability: Science, Practice and Policy. 2017; 13(1): 13-26.
31. Walker W. Information technology and energy supply. Energy Policy. 1986;14(6):466-88.
32. Cox S. Information technology: the global key to precision agriculture and sustainability. Computers and electronics in agriculture. 2002; 36(2-3):93-111.
33. TheGreenSAGE™ https://bitmantis.launchrock.com/ accessed on May 28, 2020
34. Gipta MK, Ganapuram S. Vertical Farming Using Information and Communication Technologies. https://www.infosys.com/industries/agriculture/insights/documents/vertical-farming-information-communication.pdf accessed on May 28,2020

4

Precision Agriculture

4.1 Introduction

In the modern era, agricultural productivity seems to have reached a maximum production due to the easy availability of fertilizers and pesticides/insecticide, which are used to improve crop productivity. Excess uses of these products and unawareness about field constraints can decrease the crop productivity and endanger the environmental balance in the cultivation area. Therefore, there is an urgent need to improve agricultural system and efficient utilization of these resources for profitable and environmentally sustainable agriculture[1-3].

Government has also an emphasis on effective and efficient use of available resources and farmers are also adopting more scientific approach during crop production such as remote sensing, global positioning system, and new equipment to make their agricultural operation more precise.

Precision agriculture is one of the many advanced farming approaches that make production more efficient. It is basically an approach of farm management with the help of information technology to confirm that the cultivated crops and soil receive desire input for optimum health and productivity. In other words, precision agriculture utilizes different tools and technologies to identify in-field soil and crop variability for improving farming practices and optimizing agronomic inputs[4]. To reduce dependency on excessive use of chemical inputs in agriculture, several technologies have been applied to make agricultural products safer and to lower their adverse impacts on the environment and human beings, a goal that is consistent with sustainable agriculture. Precision agriculture has emerged as a valuable component of the framework to achieve this goal[5]. Further, Pierce and Nowak[6] define the precision farming that, the application of technologies and principles to manage spatial and temporal variability associated with all aspects of agricultural production for improving crop performance and environmental quality. Precision agriculture is the combination of many different elements working together in a synchronised way with the help of Geographic Information System (GIS) as shown in Fig 4.1. Precision agriculture is basically an information technology based agricultural management system. In this modern management system,

Geography information system (GIS) based analysis is extensively used to get land and crop related data accurately and timely, which help farming community to take correct decision[7].

The fundamental goal of precision agriculture is to ensure sustainability, timely management, profitability, and protection of the environment. Many researchers and scientists have categorised precision agriculture as satellite agriculture, as-needed farming and site-specific crop management.

Precision agriculture highly depends on engineered equipment, information technology, and different software tool. These tools are in a position to provide real time information of crop, soil, climatic conditions, and weather predications.

Data analysis and predictive software are also incorporated in precision agriculture, which guide farmers about soil management, optimal planting / sowing /harvesting time, and crop rotation, etc.

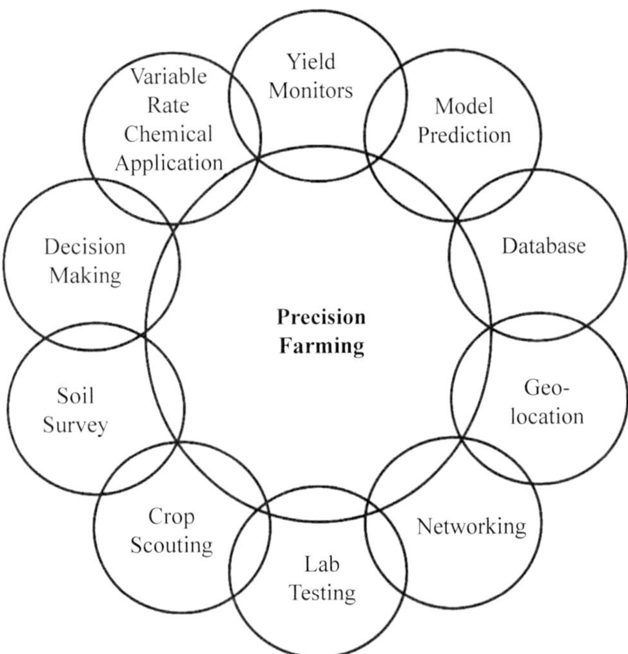

Fig. 4.1: Different Components of Precision Agriculture

The Benefits of precision farming are as follows[8]:

a) Monitor the soil and plant physicochemical parameters: by placing sensors (electrical conductivity, nitrates, temperature, evapotranspiration, radiation, leaf and soil moisture, etc.) the optimal conditions for plant growth can be achieved.

b) Obtain data in real time: the application of sensing devices in the fields will allow a continuous monitoring of the chosen parameters and will offer real time data ensuring an updated status of the field and plant parameters at all time.

c) Provide better information for management decisions.

d) Save time and costs: reduce fertilizer and chemical application costs, reduce pollution through less use of chemicals.

e) Provide better farm records essential for sale and succession.

f) Integrated with farm management software to make all activities on the farm easier and to improve farm productivity.

4.2 Precision Farming in Indian Context

Adaptation of precision farming in India, can accelerate socio-economic changes such as economic growth and energy consumption, etc. Due to poor infrastructure, social and economic conditions of the farmer, and small size landholding, it is not adapted with open heart by Indian farmers. It has been estimated that about 58 per cent of agricultural landholding in India is less than 1 ha. Only the Indian state Rajasthan, Punjab, Gujarat, and Haryana having land holding more than 4 ha. There is huge scope of implementing this farming system in Punjab and Haryana, especially for wheat and rice crops. Small landholding farmers can also adopt precision farming with the use of small farm machinery, which may run on green and renewable fuel like bio-oil, biogas, and electricity produce through crop residues. Such farming may help in sub surface drip irrigation for precise and effective use of water and fertilizer.

4.3 Components of Precision Agriculture

Precision agriculture (PA) consists of many different components and types of equipment and it is the combination of application of different technologies. The technologies and equipments are mutually inter related and are responsible for overall developments. Some of the important tools, comprehensively used in PA are discussed below:

a) Global Positioning System (GPS)

b) Geographic Information System (GIS)

c) Grid Sampling

d) Variable Rate Technology (VRT)

e) Yield Maps

f) Remote Sensing

g) Proximate Sensors

h) Computer Hardware and Software

i) Precision irrigation systems

4.3.1 Global Positioning System (GPS)

Global positioning system (GPS) receivers provide a method for determining location anywhere on the earth. Accurate, automated position tracking with GPS receiver allow farmers and agricultural service providers to automatically record data and apply variable rates of input to smaller areas within larger fields. The accuracy of obtained data/information depends of method of installation, degree of technology. location of satellites, and atmospheric conditions[9]. Borgelt et al.[10] mentioned that the use of GPS technology is an efficient and effective method of providing location data for precision agriculture applications. The higher accuracies can be obtained with carrier phase kinematic positioning methods, but this required more time and a less robust technique with a greater potential for data acquisition problems. The accuracy of GPS was estimated by Webster and Cardina[11] with circular areas of 5, 50, and 500 m^2 and measured using point and polygon features of a GPS and found that patches had errors ranging from 7 to 45 per cent, 6 to 15 per cent, and 3 to 6 per cent respectively, when compared to actual measurements. It was also reported that as patch size increased, errors decreased. Therefore, it is concluded from Webster and Cardina[11] study that larger plot size is desirable for higher accuracy.

An interactive, portable system to record field data directly into a digital database comprising yield, soil, road, water and contour maps overlain on air photos or remote sensing imagery was developed by Shanwad et al.[12]. In their portable system, GPS receiver is linked to a note book computer displaying appropriate, pre-loaded information layers, and a software package then combines incoming GPS signals with the displayed data to allow the user to see where they are with respect to the map components. Further, system also allows the user to save GPS data to view and track field activity at a later date.

4.3.2 Geographic Information System (GIS)

A Geographic Information System (GIS) is basically a tool which creates visual images of data and performs spatial analyses in order to make informed decisions. The GIS technology is combination of hardware, software, and data. The data can represent almost anything imaginable so long as it is associated with geographic component. The hardware can be anything from a desktop computer or laptop to satellites, drones, and handheld GPS units[13]. GIS can integrate all types of information and interface with other decision support tools and display analysed information in the form of maps that allow better understanding of interaction between crop yield, soil fertility, pests, weeds and other factors. Moreover, GIS is defined as it is a set of tools for collecting, storing, retrieving, transforming and displaying spatial data from the real world for a particular set of purpose[14].

GIS application in agriculture such as agricultural mapping plays a vital role in monitoring and management of soil and irrigation of any given farm land. It is an essential tool for management of agricultural sector by acquiring and implementing the accurate information into a mapping environment. It also helps in management and control of agricultural resources and helps in improvement of the current systems of getting and producing GIS agriculture and resources data.

In 2001 a field–level geographic information system (FIS) was designed by Zhang and Taylor[15] specifically for research in precision agriculture. In initial inception, they analyse the response of soybean yield to soil electric conductivity using mathematical/logic query and simple statistics functions though FIS. Later, they demonstrate the use of FIS in delineating management zones through morphological filtering and spectral filtering. This study reveals the potential of FIS in problem solving and decision making for a variety of precision–agriculture applications. Runquist et al.[16] also developed a field-level GIS (FIS) containing analytical functions for spatial data analysis in precision agricultural research.

4.3.3 Grid Sampling

Grid sample is a method of taking soil samples at a regular interval across the selected agricultural filed, and its size is selected to provide the desirable data resolution. Smaller grid is usually recommended for accurately capturing the desired information. Grid sampling reveals that how the nutrients are distributed across the field. Higher number of soil samples enable to farmers for better understanding of the nutrients availability. It allows farmers to avoid over application of fertilizer in the areas where nutrient levels are high and allow to enrich the soil with fertilizer in the areas where nutrient levels are low. Using grid sampling approach, the fertilizer prescription can be optimised, and return on investment can be improved. The grid sampling technique normally recommended when there is little information available about the variation in nutrient levels across a field. It may be useful for such a field where variability is expected but the field history is unknown, topography is uniform but differences in soil type occur. The gird sampling services are valuable effort which allow the data to be used for input for variable rate applications for optimizing seeding and fertilizer. Heermann et al.[17] establish sampling strategies, analysis techniques and decision procedure during field study at farmers filed. They reported that farmers had demonstrated the value of additional scientific information by changing their management strategy towards irrigation and fertilizer applications. The grid sampling provides the opportunities to decrease input costs and potentially increase net income. With the proper layer of information, grid sampling can improve the accuracy of variable rate application. Lund et al.[18] reported the importance of grid sampling, which was taken for producing map of soil electric conductivity. Grid sampling enable them to reduce the amount of lime applied on high pH soils.

4.3.4 The Variable Rate Technology (VRT)

The Variable Rate Technology refer to any tool or system that enables farmers to vary the rate of crop inputs for specific location. It is the prime motive of precision farming and combined with other technological advances for driving farming forward at a rapid rate of speed and growth. Using VRT in actual use, one can improve the productivity or reduce the cost of production and diminish the chance of environmental degradation caused by excess use of inputs[19]. There are huge gaps in efficiency between traditional farming, and precision farming and this gap can be bridged through utilising VRT with three basic approaches, i.e. map-based, sensor based, and manual.

Economic feasibility of VRT for Nitrogen on Corn as estimated by Thrikawala et al.[20] and reported that potential leaching losses are reduced in all experimental fields exhibiting some spatial variability since deficiencies and excesses in N application are avoided. The environmental benefits of variable-rate technology increase with field fertility variability due to the relative increases in fertilizer use efficiency as compared to a single application rate method. Babcock and Pautsch[21] also experimentally evaluated the net profitability of variable-rate technology and found that the applying variable rates would increase yield by 0.05 to 0.50 bushels per acre, and would reduce fertilizer costs by $1.19 to $6.83 per acre. The economics of VRT investment decision for agricultural sprayers was carried out by Mooney et al.[22] and found that it has high ownership costs but low recurring annual costs. Further, they also found that under a baseline scenario, VRT systems using high-resolution NDIV sensors and those using aerial NDVI imagery were found to become profitable at input saving levels of 11 per cent or above. The adoption of variable rate technology is currently estimated at 15 per cent in North America and is expected to continue to grow rapidly over the next five years.

4.3.5 Yield Mapping

Yield mapping is a technique in agriculture of using global positioning system data to analyse variables such as crop yield and moisture content in a given field. It is the basis for understanding the variability, and improving the management according to the increase in profit[23]. It was developed in the 1990s and uses a combination of global positioning system technology and physical sensors, such as speedometers, to track crop yields, grain elevator speed, and combine speed.

This data produces a yield map that can be used to compare yield distribution within the field from year to year. This allows farmers to determine areas of the field that, for example, may need to be more heavily irrigated or are not yielding any crop at all. It also allows farmers to show the effects of a change in field-management techniques, to develop nutrient strategies for their fields, and as a record of crop yield to use in securing loans or renters[24]. Two main errors have been identified in many yield maps; first is the lag time between detachment and sensing of the grain, which offsets the yield position along the route of the combine, and second is the unknown width of crop entering the header[25].

4.3.6 Remote Sensing

The remote sensing technique is extensively used to gather exhaustive information in space and time even from inaccessible areas. It is a science of making interfaces about material objects from measurement at distance without coming into physical contact (Fig. 4.2). Remote sensing is used in numerous fields, including geography, land surveying and most Earth science disciplines.

Government has also emphasised that sensor based monitoring system must be incorporated into the agricultural sector at an early stage, i.e. from growing period to until harvesting and should be updated regularly. During updating monitoring system, data regarding irrigation, crop production, conditions and yield to be incorporated, which enable early warning system. The statistically accurate data, enable stakeholder to recognise any discrimination in their farm management through an early warning system. Wireless sensor networks are another emerging technology that can provide processed real-time field data from sensors, which are physically disseminated in the field[26].

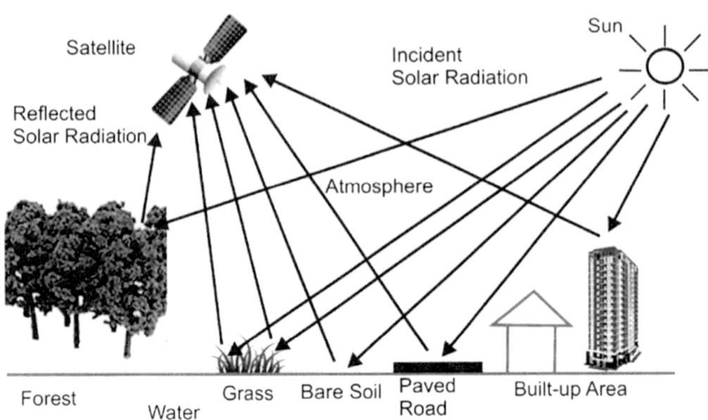

Fig. 4.2: Remote Sensing Mechanism

El Nahry et al.[27] realize the land and water use efficiency and determining the profitability of precision farming economically and environmentally through remote sensing and GIS techniques. The experimental results reveal that they save about 23.56 tonnes of fertilizer and improve natural drainage system by digging vertical holes to break down massive soil layer and to leach excessive salts.

4.3.7 Proximate Sensing

Integration of sensor based technology in precision agriculture that allows to monitor the crops parameters to obtain good quality food and increase overall crop productivity. The sensor based technology have their own importance because the impact of a factor like climate change, food shortage, poor food distribution, etc., can be reduced with the use of such technology. The sensor helps in maintaining and controlling favourable conditions for development of crops and plants[28]. A large number of sensors are used in precision farming such as location sensor which use the signal from global positioning system satellite to determine the location of growing area with latitude, longitude, and altitude. The optical sensors measure soil properties. With different light spectrum, optical sensor is capable to determine clay, organic matter present in soil, and moisture content of soil. Dielectric soil moisture sensor also used to check the level of moisture present in soil. The soil nutrient and pH level can be measured using electrochemical sensor whereas soil compaction measures through a mechanical sensor. Airflow sensors enable to measure soil air permeability. Lowenberg and Erickson[29] reported that the pH soil sensor had been potential for modest farm level economic benefits.

4.3.8 Computer Hardware and Software

Large number of data gathered through GPS, GIS, remote sensor and other useful technologies are adapted in precision agriculture. It is difficult to understand the trend and indication of raw data by the growers. Therefore, to make it useable and understandable by grower in the form of maps, graphs, and visuals or reports, computer support is desirable along with requisite software support. As it was mentioned earlier that precision farming management is combination of smart monitoring, planning and control of different agricultural processes. To manage all these varieties of software and hardware incorporated, and they have poor interoperability data exchange between ICT components, which hamper the purpose of the precision farming. To overcome from this situation Kruize et al.[30] developed a software ecosystem that addresses all these challenge issues.

4.3.9 Precision Irrigation Systems

The sensor based irrigation system enhances the efficiency of irrigation processes and minimizes water losses. Precision irrigation focuses on individual plants or small areas within a farm. It advocates the water conservation opportunity. The main objective of the precision irrigation is to optimize water application efficiency based on real climatic data collected at the farm. Precision irrigation scheduling refers to technologies that help growers determine more precisely when crops need to be watered and how much water they require. With smart irrigation scheduling, growers are able to use their water more efficiently, either by reducing or by keeping constant across the amount of applied water, while maintaining or improving yields. Having more precise knowledge of soil moisture levels also has a number of peripheral benefits, such as pest control.

References

1. Donner SD, Kucharik CJ. Corn-based ethanol production compromises goal of reducing nitrogen export by the Mississippi River. Proc. Natl. Acad. Sci. 2008; 105: 4513–4518.
2. Zhang X, Davidson EA, Mauzerall DL, Searchinger TD, Dumas P, Shen Y. Managing nitrogen for sustainable development. Nature 2015; 528: 51–59.
3. Zillén L, Conley DJ, Andrén T, Andrén E, Björck S. Past occurrences of hypoxia in the Baltic Sea and the role of climate variability, environmental change and human impact. Earth-Sci. Rev. 2008; 91: 77–92.
4. Khanal S. Fulton,J. Shearer S. An overview of current and potential applications of thermal remote sensing in precision agriculture. Computers and Electronics in Agriculture 2017; 139: 22-32.
5. Liaghat S, Balasundram SK. A review: The role of remote sensing in precision agriculture. American journal of agricultural and biological sciences 2010; 5(1): 50-55.
6. Pierce FJ, Nowak P. Aspects of precision agriculture. Adv Agron 1999; 67:1–85
7. Ye J, Chen B, Liu Q, Fang Y. A precision agriculture management system based on Internet of Things and WebGIS. In 21st International Conference on Geoinformatics (pp. 1-5). IEEE 2013.

8. Folnovic T Benefits of Using Precision Farming: Producing More with Less. https://blog.agrivi.com/post/benefits-of-using-precision-farming-producing-more-with-less (retrieve on 17.03.2020)
9. Pfost DL, Casady WW, Shannon K. Precision agriculture: global positioning system (GPS). Extension publications (MU) 1998.
10. Borgelt SC, Harrison JD, Sudduth KA, Birrell SJ. Evaluation of GPS for applications in precision agriculture. Applied Engineering in Agriculture 1996; 12(6): 633-638.
11. Webster TM, Cardina J. Accuracy of a global positioning system (GPS) for weed mapping. Weed Technology, 1997;11(4: 782-786.
12. Shanwad UK, Patil VC, Dasog GS, Mansur CP, Shashidhar KC. Global positioning system (GPS) in precision agriculture. In The Asian GPS Conference 2002: 24-25.
13. Dornich K. Use of GIS in Agriculture. https://smallfarms.cornell.edu/2017/04/use-of-gis/ (accessed on April, 29, 2020)
14. Burrough PA. Principles of Geographical Information System for Land Resources Assessment. Clarendon Press California 1986.
15. Zhang N, Taylor RK. Applications of a Field–Level Geographic Information System (FIS) In Precision Agriculture. Applied engineering in agriculture 2001; 17(6): 885-892.
16. Runquist S, Zhang N, Taylor R. Development a field-level geographic information system. Computers and Electronics in Agriculture 2001; 31: 201-/209.
17. Heermann DF, Hoeting J, Thompson SE, Duke HR, Westfall DG, Buchleiter GW, Westra P, Peairs FB, Fleming K. Interdisciplinary irrigated precision farming research. Precision Agriculture. 2002; 1;3(1):47-61.
18. Lund ED, Colin PE, Christy D, Drummond PE. Applying soil electrical conductivity technology to precision agriculture. In Proceedings of the fourth international conference on precision agriculture. Madison, WI, USA: American Society of Agronomy, Crop Science Society of America, Soil Science Society of America. 1999: 1089-1100.
19. Sharma S, Manhas SS, Sharma RM, Lohan SK. (2014). potential of variable rate application technology in India. Agricultural Mechanization in Asia Africa and Latin America, 2014; 45(4): 74-81.
20. Thrikawala S, Weersink A, Fox G, Kachanoski G. Economic Feasibility of Variable-Rate Technology for Nitrogen on Corn. American Journal of Agricultural Economics 1999; 81(4): 914-927.
21. Babcock BA, Pautsch GR. Moving from uniform to variable fertilizer rates on Iowa corn: Effects on rates and returns. Journal of Agricultural and Resource Economics, 1998; 23(2):385-400.
22. Mooney DF, Larson JA, Roberts RK, English BC. Economics of the Variable Rate Technology Investment Decision for Agricultural Sprayers. presentation at the Southern Agricultural Economics Association Annual Meeting, Atlanta, Georgia 2009: No. 1369-2016-108610.
23. Wang MH. Field Information Collection and Process Technology. Agriculture Mechanization 1999; 7: 22-24
24. Franzen DW, Casey FXM, Derby N. *Yield Mapping and Use of Yield Map Data* (PDF). *University of North Dakota*. Retrieved April 30, 2020.
25. Blackmore BS, Marshall CJ. Yield mapping; errors and algorithms. In Proceedings of the Third International Conference on Precision Agriculture. Madison, WI, USA: American Society of Agronomy, Crop Science Society of America, Soil Science Society of America 1996: 403-415.
26. Camilli A, Cugnasca CE, Saraiva AM, Hirakawa AR, Corrêa PL. From wireless sensors to field mapping: Anatomy of an application for precision agriculture. Computers and Electronics in Agriculture. 2007; 58 :25-36.

27. El Nahry AH, Ali RR, El Baroudy AA. An approach for precision farming under pivot irrigation system using remote sensing and GIS techniques. Agricultural Water Management. 2011; 98(4): 517-31.
28. Padilla-Medina JA, Contreras-Medina LM, Gavilán MU, Millan-Almaraz JR, Alvaro JE. Sensors in Precision Agriculture for the Monitoring of Plant Development and Improvement of Food Production. Journal of Sensors. 2019: Article ID 7138720
29. Lowenberg-DeBoer J, Erickson B. The economics of direct soil sensing in agriculture. The Second Global Workshop on Proximal Soil Sensing – Montreal 2011
30. Kruize JW, Wolfert J, Scholten H, Verdouw CN, Kassahun A, Beulens AJ. A reference architecture for Farm Software Ecosystems. Computers and Electronics in Agriculture. 2016; 1: 125:12-28.

5

Hydroponic Agriculture

5.1 Introduction

Hydroponics is a method of growing plants in a nutrient solution rather than in soil. Terrestrial plants may be grown with only their roots exposed to the nutritious liquid.

Traditionally, the soil supports the crop's roots by helping them remain upright and ensuring the delivery of water and essential nutrients. In hydroponics farming, crops are supported by an inert medium such as perlite, gravel, or other substrates. Therefore, the nutrients are supplied by using various practices that bring mineral nutrient solutions to the crops. In the present situation there excessive use of chemical fertilizers, affecting fertility of soil is badly affected. The fertility of the soil is falling that affects the plants growth but in hydroponics, the plant remains unaffected. Plants grow faster than in soil. Produce better quality of plants and higher yields. It use less water. It is possible to use hydroponics on outdoor crops, but most of the hydroponics production is in greenhouse. Hydroponic farming system has the potential to reduce the problem associated with traditional crop production. Therefore, it could serve to be a turning point in feeding the global population. The greenhouse and its environment control system are the same whether plants are grown conventionally or with hydroponics. The difference comes from the support system and the method of supplying water and nutrients.

The hydroponic farming has been successfully demonstrated by the commercial growers for fast-growing crops such as tomatoes, cucumbers, spinaches, lettuce, radishes, strawberries, bens, spring onions, coriander, and ornamentals. Indeed, hydroponics technology enables a more precise control of micro climatic conditions. In this technique, the vigorous development of a root system is taking place, which enable the uptake of essential nutrients from nutrient solution for better crop yield. The initial investment in hydroponics farming is high and required constant energy input. Owing to this, hydroponics farming is recommended for growing high value crops in developed countries such as the United States, Canada, Europe and Japan.

In the past two decades, it has been observed that the adaptation of hydroponics farming in India is steadily increasing. Such farming system is becoming more viable growing option in the present context where the dependency on water and pesticides is at an all-time high. It has led traditional agriculture cultivation to become deliberately unsafe and unsustainable. Adoption of hydroponics farming is continuously increasing in India due to climatic variation, poor availability of fertile soil, and clean and surplus water for irrigation. Many studies reveal that hydroponic farming is economically feasible, and it serves as possible solution to increasing population. Indian farmers are facing so many challenging issues like year-round production, drought conditions, and global warming. Hydroponics farming is capable of resolve these issues and strengthen local food grower to produce food for India's larger population. In India, hydroponics farming is becoming popular among garden lovers who usually grows their own green vegetable under greenhouse in their rooftop. Smart farming is the future of agriculture for the production of good quality crops using intelligent sensor to control growth parameters. Hydroponics farming system integrated with automated sensor and exact interface in Bayesian Network (BN) increases the crop yield by 66.67 per cent than the manual control[1].

5.2 Crops Growing in Hydroponic System

There are a variety of plants that can grow indoors, without needing too much sunlight. Some of the edible easy-to-grow plants that work very well with the Hydroponic cultivation are listed in Table 5.1.

Table 5.1: Plants Commonly Grown in Hydroponic System

S. No.	Crop	Description
1.	Tomatoes 	Variety of tomatoes including traditional and cherry ones are widely grown by hydroponic growers. Tomatoes require much light therefore, grow lights are essentially required to grow indoors.
2.	Lettuces 	Lettuce is the easiest vegetable to grow hydroponically. It is the perfect ingredient for the salad sandwich. Lettuces can be grown in any Hydroponics system, including the NFT, Aeroponics, Ebb & Flow, etc.

S. No.	Crop	Description
3.	Cucumber 	Cucumbers flourish in hydroponics because of their fast growth rate and needs for warmth, moisture, and nutrients which are effectively filled in the hydroponic system. Cucumbers are one of the highest yielding plants normally grown in a hydroponic system.
4.	Radish 	Radishes are a rapidly maturing cool weather crop. They are among the easiest vegetables to grow hydroponically. It's better to start from seeds, seedlings prepared within 3 - 7 days. Radishes thrive in cool temperatures and do not need any lights.
5.	Spinach 	Spinach is one of the mostly grown plants in hydroponic system. To grow the spinach well, submerge the roots in the nutrient solution. Enjoy more advantages by growing spinach hydroponically such as less pests, more crops and less space required.
6.	Kale 	Kale's popularity can bring in a good profit for farmers. The crop grows relatively quickly with a six-week turn from transplant to harvest or can be harvested partially to regrow. It is a great vegetable for a healthy person with proven health benefits. Deep water culture and nutrient film technique type hydroponic system mostly recommended for growing kale.

S. No.	Crop	Description
7.	Beans 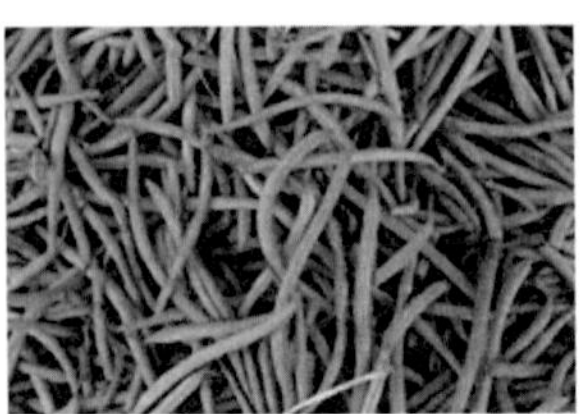	One of the most productive and low-maintenance vegetables that can be grown hydroponically. Seed germination usually takes 3 - 8 days. Harvesting begins after 6 - 8 weeks. After that, one can continue the crop for 3 - 4 months.
8.	Strawberries 	This is an easy plant to grow, and they do really well in hydroponic systems. One of the major benefits of growing strawberries hydroponically, aside from the magnificent taste, is that they can be grown at an elevated height.
9.	Spring onions 	Spring onions are very young onions harvested before the bulb gets to swell and grow. One pot can sprout dozens of onions and be harvested every 3 or 4 weeks. Spring onions are loaded with vitamins C and K which are both essential for healthy bones. Spring onions natural properties are most commonly used to treat viral infections such as flu and colds. They also contain vitamins B and A.
10.	Coriander 	Coriander can be raised in a range of hydroponic systems, growing coriander in hydroponic NFT system is most common, DFT and DWC are the ideal systems. Temperature of at least 20 °C is required for growing this herb with moderate light level for seedlings and increasing light level as plants mature.

5.3 Components of Hydroponic System

5.3.1 Growing Systems

There are a number of materials that can be used as growing media in hydroponics, and all have different properties than another type of media. The most widely growing media used in hydroponics system are Rockwool, Expanded clay aggregate, Coconut Fiber/Coconut chips, and Perlite or vermiculite. It is difficult to identify the best growing media suited for all situations. Rockwool is a sterile, porous, and non-degradable medium that is composed primarily of granite and/or limestone. Perlite is mainly composed of minerals that are subjected to very high heat, which then expand it like popcorn so it becomes a very light weight, porous and absorbent. It has a neutral pH, excellent wicking action, and is very porous. Lightweight expanded clay aggregate is pelletized clay and fired in kilns until it expands into small, orangey-red balls, and often called grow rocks[2].

5.3.2 Sand/Stone Culture

Sand can be used as a growing medium, but it has poor aeration qualities. The growing media for hydroponics system should possess good water holding capacity combined with good drainage qualities, and allow the roots to feed, exhaust CO_2, and ingest oxygen properly. The gravel culture system is an easy and inexpensive way to grow plants hydroponically. In this growing technique, plant consists of a deep bed (18 -24 inches) of sand, pea stone or trap rock placed in a plastic lined trough or bed, which slopes to one point to drain off excess nutrient solution as shown in Fig 5.1. Seedlings are set directly into this medium and watered several times per day with the nutrient solution.

Fig. 5.1: Sand and Gravel Culture Hydroponic System

5.3.3 Troughs and Pipes

Trough and pipe cultivation is becoming popular among soilless crop production and hydroponics because it gives the grower more control over the growing media. Open and closed plastic troughs or PVC pipe, and grow bags are the most economical hydroponics system. These may contain just the nutrient solution or may be filled with growing media. Usually, these trough and pipes are mounted on rollers or on movable racks with suitable spacing as shown in Fig. 5.2. In the PVC pipe, there is a detachable lid for easy cleaning and end caps for draining the waste nutrient solutions.

Fig 5.2: PVC Pipe for Hydroponic System

5.3.4 Trays

A tray is a container specifically designed to hold one or more plants in a hydroponics growing system. Depending on the type of hydroponics system in use, these trays may have to leach valves to allow water to drain out of the growing medium when necessary. In general, trays are long, rectangular, and shallow with room to accommodate root structures and growing plants and are mostly used for hydroponics gardens than other types of indoor gardens as shown in Fig. 5.3. Trays are made from moulded plastic or waterproof plywood.

Fig. 5.3: Growing Tray

5.3.5 Bags

Hydroponics grows bags are essentially made off UV stabilized polyethylene film and filled with growing media (Fig 5.4). Grow Bags are used as a component of food production for variety of crops. The bags are laid end to end and drip tube or soaker hose inserted to supply the nutrients. The bags may be good for several crops before they have to be replaced.

Fig. 5.4: Growing Bags

5.4 Type of Hydroponic Farming

There are two main types of hydroponic farming systems:

1) Passive system
2) Active system

5.4.1 Passive Hydroponic System

There are many types of passive hydroponics systems. In a passive hydroponics system, plant is suspended over the nutrient solution and relies on its roots to access it. First, passive hydroponics system is the Kratky Method, which employs plants suspended into the nutrient tank. The second method known as wicking works just like capillary action in the natural world of plants but allows an oxygen zone between growth media and nutrient solution.

The Kratky Method was discovered by B.A. Kratky from the University of Hawaii. It can be seen as the Deep Water Culture without using a pump. This method is purely a completely passive system. There is no electricity used as well as no pumps and wick needed. Growers also don't have to change nutrients in the reservoir much often like other systems. Krafty is a low-maintenance system that can work on its own for weeks.

To reach full potential growth, plant requires oxygen, water, lighting to survive, macro nutrients, and micronutrients. All these essential things are easily supplied in Kratky passive system by adding nutrients to the container or reservoir, placing plant into net pot with growing media, submerged plants roots partly into water and partly exposed to the air. As the plant growth increases, water level reduce and leaving a gap of the roots exposing to the air as shown in Fig. 5.5. This air gap is essential because it is where the plants respire.

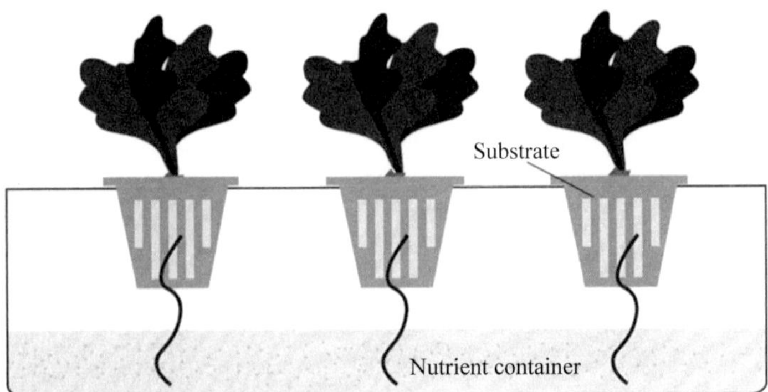

Fig. 5.5: Passive Hydroponic System

5.4.2 Active Hydroponics System

Active hydroponics is the plant growing technique, where the roots of the plant are fully submerged in nutrient solution, and an air pump produces air bubbles or oxygen for the plant roots. The ebb and flow method also uses an active system by temporarily flooding the roots in the growing tray with a pump. The pump is submerged into the nutritive solution, when it on, it pumps the nutrients up to the plant tray and floods the roots for feeding. After running the pump for about 20 or 30 minutes, switch off the pump and the excess solution is allowed to drain slowly back down into the reservoir or recovery. The ebb and flow system is usually automated with a water pump on a timer that floods the growing trays at specific intervals as shown in Fig 5.6.

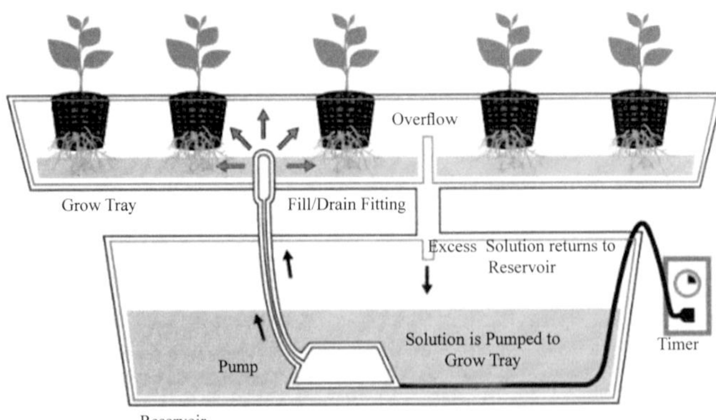

Fig. 5.6: Ebb and Flow System

Apart from the passive and active hydroponics system, there are few popular hydroponics farming systems are as follows:

5.4.2.1 Deep Water Culture Hydroponic System

This is the simplest **method of plant production by means of suspending the plant roots in a solution of nutrient-rich, oxygenated water** and also known as raft/pond or float systems as shown in Fig. 5.7. In this method, floating rafts suspend plant roots into a pond of water often 8-12 inches deep. Compared to other hydroponic techniques, such as Nutrient film technique, this technique is relatively low-cost to set up and can be easily reproduced by a home grower. Since there is a relatively large reservoir of nutrient rich water for each plant, there is buffering for pH, EC, and temperature, meaning those elements of the systems won't fluctuate as fast as they might in a Nutrient film technique system. It is a most popular method for at-home gardeners. This technique is effective and extremely easy hydroponic system to assemble and maintain. Concentration of nutrient solution should maintain properly; high concentration of nutrient solution can fry roots causing nutrient lockout.

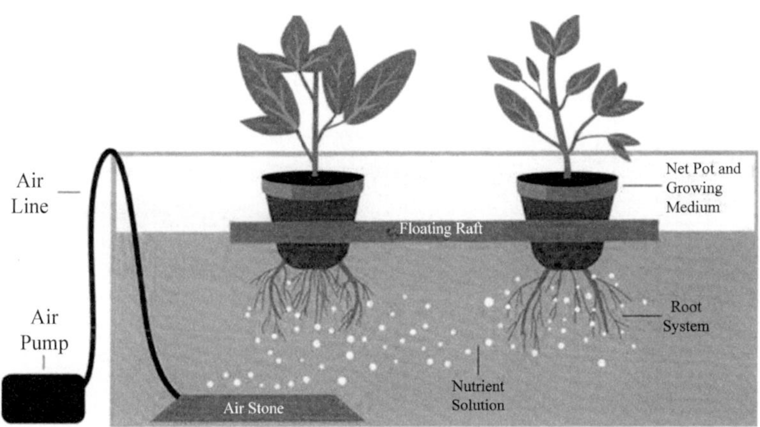

Fig. 5.7: Deep Water Culture Hydroponic System

5.4.2.2 Nutrient Film Technique System

This is the kind of hydroponic technique where in a very shallow stream of water containing all the dissolved nutrients required for plant growth is pumped into the growing tray (usually a tube) and flows over the roots of plant, and then drain back into the reservoir. As there is no growing media, the plants are supported in a small plastic basket with the roots dangling into the nutrient solution as shown in Fig. 5.8. Such plants growing techniques are very susceptible to power outage and pump failure. If the flow of nutrient solution is interrupted, plant roots dry out rapidly. The nutrient film technique is often used to grow smaller and quick growing plants like different types of lettuce. Apart from lettuce, commercial growers also use this system to grow herbs, baby greens and strawberries. This technique is a versatile and familiar hydroponics system that include components the same as ebb and flow but different in configuration.

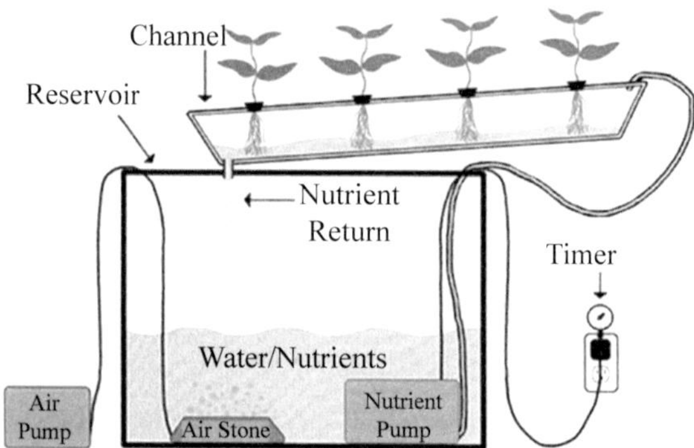

Fig. 5.8: Nutrient Film Hydroponic Technique

5.4.2.3 Aeroponic System

In an aeroponics system, plants are grown in troughs, tubes or other types of chambers. The roots of plant in this method are not in any kind of solid material such as rockwool or soil. Normally, roots hang in space and are periodically bathed in the nutrient mist as shown in Fig. 5.9. This method enables a plant to grow larger and more easily absorbs nutrients and oxygen.

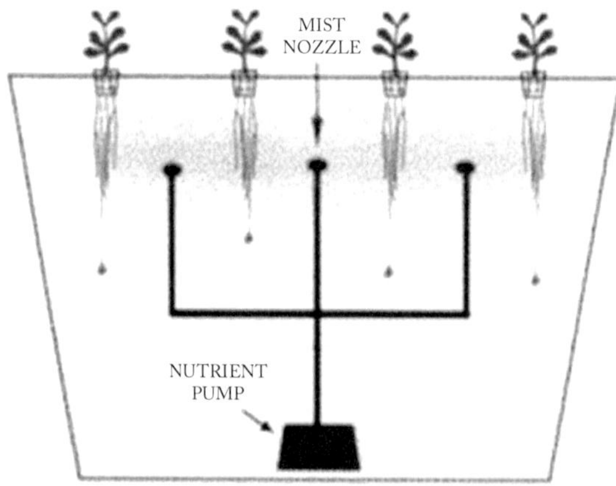

Fig. 5.9: Aeroponic System

5.4.2.4 Drip Hydroponic System

In drip hydroponic system, the nutrient solutions and water are pumped from a reservoir to each plant using a small drip line as shown in Fig. 5.10. The nutrient rich water flows over the growing media and drip down over the plant roots. Drip hydroponic system may be of recovery or a non-recovery type. In non-recovery methods, nutrient water that runs over the plant's roots doesn't return to a reservoir whereas, in recovery method nutrient water return back to a reservoir.

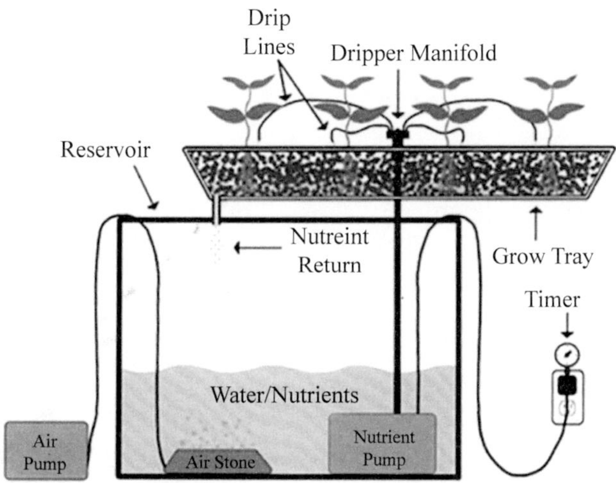

Fig 5.10: Drip Hydroponic system

5.5 Installation of Homemade Hydroponic System

The hydroponics growing system at home level can easily be installed. First identify the location where it is to be installed such as greenhouse, basement of the house or on the rooftop. Selected floor should be levelled to ensure proper flow of water and nutrient. If the system is installed outside the house, care should be taken for wind velocity, water level, and solar radiation. If it installed inside the house supplement grow light is desirable to maintain the photosynthesis reaction. Assembly of homemade hydroponics farming system is completed in following steps:

Step 1: Assemble the PVC Pipe

Normally, 6 inches PVC pipe is found suitable for home hydroponics farming. Assembly comprises with PVC pipe, stand, nutrient tank, pump and a manifold. Each growing PVC pipe has a drain pipe which drain nutrient solution back to tank. The manifold assembles on top of the PVC pipe sends pressure water to the tube (Fig. 5.11).

Fig. 5.11: PVC Pipes for Home Hydroponic Farming

Step 2: Mix the Nutrients and Water in the Tank

Nutrient solution to Hydroponic is just like fertilizers to soil. Hydroponics nutrient solution is a liquid filled with all the necessary nutrients so that plant roots can come into contact for its growth. The base of fertilizer prepares to use nitrogen, phosphorus, calcium, etc. in clean water. Micronutrient such as boric acids, chlorine, manganese, iron etc. are to be mixed in fertilizer base. After preparing solution adjust pH and EC level of solution. The pH level should be in the range of 5.5 to 6.5. If it's more than 7.0, then it's too alkaline, add vinegar to the mixture to reduce the pH level. If the pH value is less than 5.5, then the solution is too acidic, add baking soda to adjust the pH level. The EC level of solution must be between 0.8 to 3.0. In most cases, 1.5 to 2.5 are more appropriate. If the EC is too high, add more water to bring the EC down.

Step 3: Add Plants to the Growing Tubes

Healthy plant seedling can be transplanted in hydroponics garden, especially if unable to grow the seeds by grower. Before transplanting healthiest plants into pipe, clean all of the soil off their roots. To wash the dirt off the roots, submerge the root ball in a bucket of lukewarm to cool water as shown in Fig. 5.12. Gently separate the roots to get the soil out and if soil is left on the roots it could clog up the tiny spray holes in the nutrient tubes.

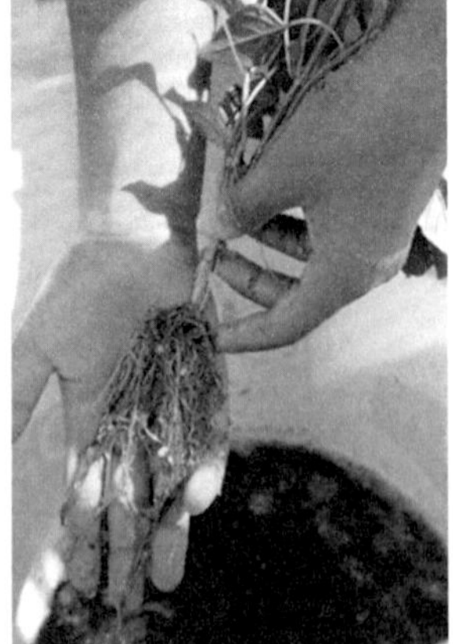

Fig. 5.12: Washing of Seedling Roots

After the roots are clean, pull as many roots as possible through the bottom of the planting cup and then add expanded clay pebbles to hold the plant in place and upright. The expanded clay pebbles are hard, but they're also very light so that they don't damage the plant roots.

Step 4: Supports & Ties

Healthy plants producing large, heavy yields have been known to snap under their own weight. With plant stakes and plant ties, this problem is eliminated. Utilize the plant clips and string to attach the plants to the trellis. The string will give them support to climb straight up, which amplifies the space in this limited zone.

Step 5: Monitor Plant Growth

The monitoring of plant crop growth during developmental stages is an important aspect of agricultural management. It enables the farmer to implement timely interventions that ensure optimal yields at the end of the season. A few weeks after planting, the plants will completely cover the trellis because they'll have all the water and nutrients they need to grow quickly as shown in Fig. 5.13. It's important to keep a close eye on plant growth and tie or cliped the plant stalks every few days.

Fig. 5.13: Healthy Plants in Hydroponic Farming

Step 5: Inspect for Pests and Diseases

Insect and mite pests are one of the biggest issues in hydroponics cultivation practices. Growers properly maintain desire light, temperatures, provide water and nutrients plants need to thrive. These conditions also set the stage for pests to survive and thrive in a nice environment with lots of healthy plants to feed on. Diseased plant can swiftly infect all the other ones since they are so close to each other. It is recommended to remove the sick plants immediately. Because plants grown hydroponically don't have to spend their energy trying to find food, they can spend more time growing. This helps them to be healthier and stronger because they can use some of that energy to fight off diseases. Since the leaves of the plants never get wet unless it rains, they're much less likely to get leaf fungus, mildew and mold. Prevention is the best way to manage pests and diseases in the hydroponics system. Diligent scouting will help in identify what pests attacked in hydroponics system. The two most reliable methods of scouting include sticky card and plant inspections. Sticky cards placed among the crop trap insects just like fly paper (Fig. 5.14).

Fig. 5.14: Sticky Cards for Pest Control

5.6 Case Studies

Hydroponics farming is setting up roots all across India. Progressive farmer Mr. Arvind Dhakad, Ratlam Dirstcit of Indian State Madhay Pradesh started cultivating strawberries (Fig. 5.15) in 3000 sq.ft since 2015. He set up the hydroponics system that is now giving him a high yield, almost 50 per cent faster growth, and his profits have increased up to 70 per cent.

Fig. 5.15: Mr. Arvind Dhakad, a farmer in Ratlam

Further, Dhakad mentioned that the minimum space need is 2×3 square feet. If using vertical hydroponics, then 30 pots can thrive, and cost about Rs 3,000 to set up. The sunlight, water, and oxygen are needed to kickstart the edible garden in the balcony, on the rooftop, windowsills, or even backyard. Moreover, he also found that, compared to conventional farming, one reduces almost 90 per cent of water usage due to water recirculation in hydroponics. Nutrient-rich matter like coco peat or pebbles acts as a medium to hold the plants in place. In vertical hydroponics farming, one bhiga land (5-8 acre) Dhakad accommodated around 70,000 strawberry plants, as against 10,000 in the conventional method (Fig. 5.16). There is very little maintenance in this method. However, if the nutrient or water requirement is not balanced, plants can die quickly[3].

Fig. 5.16: Strawberry Vertical Hydroponics Farming

5.7 Advantages and Disadvantages of Hydroponic Agriculture

Advantages

1. No soils requirement.
2. Better use of space.
3. Greater plant density.
4. Year round production regardless of the season.
5. Water-saving.
6. Unconsumed water and fertilizer can be recycled.
7. Effective use of nutrients.

8. pH control of the solution.
9. Better growth rate.
10. No weeds.
11. Fewer pests & diseases.
12. Less use of insecticide, and herbicides.
13. Labour and time savers
14. Minimal ecological and environmental pollution.

Disadvantages

1. Experiences and technical knowledge needed.
2. Water and electricity risks.
3. Possibility of system failure.
4. High Initial capital investment.
5. Long return per investment.
6. Quick spread of diseases and pests.

References

1. Alipio MI, Cruz AE, Doria JD, Fruto RM. A smart hydroponics farming system using exact inference in Bayesian network. In2017 IEEE 6th Global Conference on Consumer Electronics (GCCE) 2017: 1-5.
2. Growpackage Eco Farm. What's the Best Growing Medium for Hydroponics. Available on https://medium.com/ accessed on June 13, 2020)
3. The Better India. https://www.thebetterindia.com/230191/madhya-pradesh-how-to-grow-organic-strawberries-tips-cost-profit-home-garden-hydroponics-india-gop94/ accessed om June 18, 2020

6

Aquaponics

6.1 Introduction

Aquaponics represents the relationship between water, aquatic life, bacteria, nutrient dynamics and plants, which grow together in waterways all over the world. Taking cues from nature, aquaponics harnesses the power of bio-integrating these individual components: exchanging the waste by-product from the fish as a food for the bacteria, to be converted into a perfect fertilizer for the plants, to return the water in a clean and safe form to the fish. Just like mother nature does in every aquatic ecosystem[1].

Aquaponics is basically a combination of aquaculture and hydroponics plant production in a close-loop water system where both vegetables and fish grow simultaneously[2]. This is a closed-loop system and considered as highly efficient, produce more food compared to the traditional farming system. Importantly, it has virtually no waste and with double purpose, aquaponics farming might be an economical viable solution to the low provision of fish and vegetable in some countries. In many cases, aquaponics is advertised as a form of sustainable agriculture because it is water efficient, and has fewer environmental impacts than some forms of aquaculture[3]. That is why, aquaponics is considered as super sustainable and highly efficient because of the healthy symbiotic relationships among the plants, fish and nitrogen-fixing bacteria as illustrated in Fig 6.1. As fish excrete into the water, the water is then pumped up to the plant beds, and bacteria break down the fish excrement so the plants can absorb it.

Aquaponics farming is quite different than hydroponics farming. As discussed in previous chapter, hydroponics is the process of growing plants in a soil-less medium where nutrients are added to the water for the plants to absorb. Such systems are not as sustainable and viable as aquaponics because they require the addition of other supplemental nutrients to get desired plants growth. The quality of water decline with time and require proper disposal otherwise, it will contaminate water streams and other water sources. On the other hand, aquaponics incorporates fish to provide the natural fertilizers and nutrients plants need. This simple addition, eliminates the need to add "food" for plants. Best of all, aquaponics has a built in self-cleaning and filtration system that returns clean water to the fish. Aquaponics farming system can be more productive and economically feasible where land and water are limited. No doubt, aquaponics is complicated and requires considerable start-up costs. The production of fish and vegetables is the most visible output of aquaponics farming units. It is essential to understand that aquaponics is the management of a complete ecosystem that includes three major groups of organisms: fish, plants and bacteria.

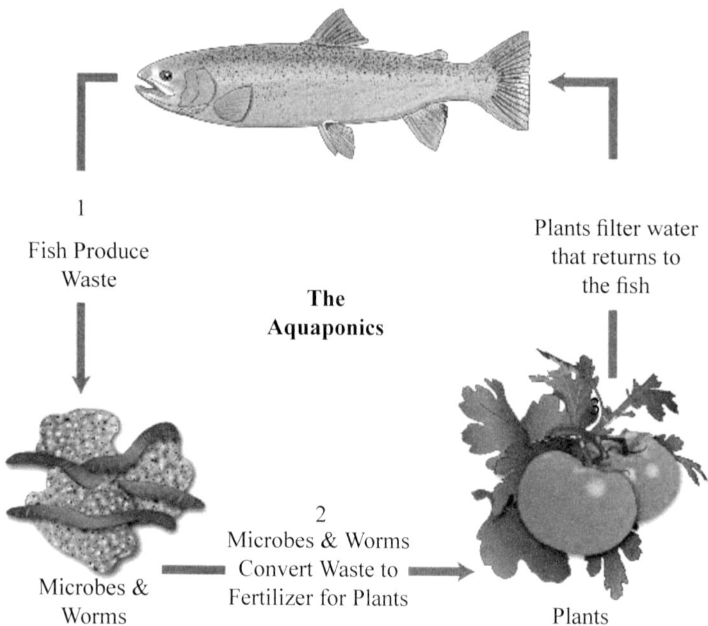

Fig. 6.1: Sustainable Aquaponics System[4]

6.2 Brief History of Aquaponics

The term Aquaponics that was coined in the 1970s and is often attributed to the rigorous works of the New Alchemy Institute and the works of Dr. Mark McMurtry at the North Carolina State University. It is difficult to pinpoint the exact time that aquaponics was first used crop production and has a widespread debate about the true origin of the aquaponics system. The ancient Aztec people who lived in Central Mexico developed a system known as Chinampas at about 1000 AD. Aztec people raised plant on stationary, or sometime movable on artificial island in shallow lakes as shown in Fig. 6.2. They did not have sufficient land to grow their food because they inhabited land that was on the shores of Lake Tenochtitlan; a fresh water lake surrounded by marshes and rising hills. The chinampas rafts made of reeds covered by soil and planted vegetable crops on them[5].

Fig. 6.2: Aztecs Raised Plants on Rafts on the Surface of a Lake

Actual research on aquaponics started in the 1970's and is continuing with many universities across the world for refining technologies to improve output. This farming system has gained momentum over the past few decades. In the early 2000's large commercial aquaponics, operations were implemented and in-depth research into their productivity was undertaken. Many researchers and growers believe that this farming system is emerging as one of the most profitable industry and commercial grower can start farming it on larger or small scale.

The online study using a chain sampling method was to document and analyse the production methods, experiences, motivations and demographics of aquaponics practitioners in the United States and internationally. It was found that about 52 per cent respondents had three years or less of aquaponics experience and typically raised tilapia or ornamental fish along the variety of leafy green vegetables, herbs and fruiting crops for their own food, for environmental sustainability and for personal health benefit[6].

6.3 Classification of Aquaponics Farming

There are three basic types of aquaponics system designs i.e., (i) Nutrient Film Technique (NFT), (ii) Media Bed, and (iii) Deep Water Culture (DWC). Before deciding the type of system to build, it is essential to know the pros and cons of each design in order to determine what fits best to fulfil the grower needs and capacity. Moreover, also think on availability of space, type of crop, annual, seasonal and daily temperatures fluctuations and importantly technical capabilities.

6.3.1 Nutrient Film Technique (NFT)

Nutrient film technique farming systems are popular because of effective space utilization and lower labour costs. In this farming technique, crops can also be grown on a vertical plane and are easily accessible, monitored and harvestable. It is found most suitable for leafy greens. In this farming system, nutrient rich water is pumped into PVC pipes as shown in Fig. 6.3. It is quite possible to grow a plant in small cups, which allow roots to absorb nutrient and water. This farming system is not found suitable for larger fruiting plants because of their weight, and they have large root mass, which may clog the water flow channel. Water flowing in pipe resembles a thin film and a biological filter is desirable to allow the beneficial bacteria to develop and convert the fish wastes into essential plant nutrients. Solid filtration is also an integral part of system to separate fish sold waste from water before circulating into pipes.

Fig. 6.3: Nutrient Film Technique Farming Systems

Mohapatra et al[7], designed and developed a portable nutrient film technique (NFT) aquaponics system and rigorously examined for 90 days to access the efficiency of the developed system. The developed technique essentially consisting of four major components i.e., (i) Fibreglass Reinforced Plastic (FRP) Round Fish Culture Tank, (ii) Biofilter unit made up of Polypropylene (PP), (iii) FRP Rectangular Hydroponics Tank, and (iv) High-Density Polyethylene (HDPE) sump. To assess the performance of developed system in actual use, 54 numbers of fish fry/m^3 was stocked (153.7 g/m^3) of pangas (Pangasius hypophthalmus) in culture tank and 27 marigold (Tagetes erecta) plants/m^2 in hydroponics tank, and found gain in weight of fish was by 397.2

per cent from initial, respectively, and marigold plant harvested 107 numbers of flowers/m^2 where the Total Ammoniacal Nitrogen (TAN) reduction in biofilter was reported about 61.97 per cent. It was also concluded that system made of lightweight FRP material was much more convenient for transportation to various remote locations and considered eco-friendly. It was remarkably pointed out that in Indian socio-economic conditions, it would be greatly benefited from the livelihood improvement of small and medium-scale farmers as well as the people in remote areas by the production of healthy food for the local market.

6.3.2 Media Filled Beds

The media bed form of aquaponics uses containers filled with rock media or similar media such as gravel or expanded clay to support the roots of plants. The bed is flooded and drained with nutrient rich water to give the plants the nutrients and oxygen they need. The media used to support the plants acts as both as mechanical and biofilter to capture and breakdown wastes. Lightweight Expanded Clay Aggregate (LECA) are balls of clay which are been shown in Fig. 6.4. It is one of the most popular media used in both hydroponics and aquaponics. This technique is best situated for backyard gardeners where larger crops can be grown. It is simple and inexpensive, and importantly media acts as filter.

Fig. 6.4: Media Bed for Aquaponics System

6.3.3 Deep Water Culture (DWC)

Deep water culture method also known as raft or float systems, uses nutrient-rich water being circulated through long water canals at a depth of about 20 cm while rafts float on top. The plant roots float directly into a pool. Plants are supported within holes in the rafts by net pots as shown in Fig. 6.5. Since there is no media to capture and process the solid wastes, filtration techniques essentially desirable in the design. This method is the most common for large commercial aquaponics growing one specific crop (typically lettuce, salad leaves or basil) and having high stocking density of fish (up to 10 and 20 kg of fish per cubic meter of the fish tank). However, it can be adapted to a low stocking density of fish production. Deep water culture farming is best suited for warmer climates because it would resist daily temperature swings, although heating the water in colder climates is costly. In addition, larger root zone plants can be used and removing plants is much easier than in media beds.

Fig. 6.5: Deep Water Culture Aquaponics

6.4 Fishes Best Fit for Aquaponics

There are variety of fishes that can be raised in an aquaponics system. There are a lot of options to choose from – some large, some smaller, some edible and some decorative. It is usually recommended that locally available fish found best suitable for aquaponics and is easier to get. However, many species of fish can rise in the aquaponics system, some need special care, whereas some are very easy to maintain. Selected species of fish commonly grown in this farming system are listed in Table 6.1.

Table 6.1: Types of Fish for Aquaponics[8]

S.N.	Species	Requirements	Advantages	Disadvantages
1.	Tilapia	• Temperature range: 28 – 30 °C • Time to one pound: 6 – 8 months • pH range: 6.5 – 9.0	• Fast growth rate • Excellent food conversion rates (1.7).	• Breed very quickly; this can be an issue if one has a small aquaponics system.
2.	Murray Cod	• Temperature range: 08 – 25 °C • Time to one pound: 12 – 18 months • pH range: 7 – 8	• It can live for a long time. • Relatively hard and can accommodate fluctuations in temp and pH.	• With increase in size they become carnivorous to other smaller fish.
3.	Catfish	• Temperature range: 24 – 30 °C • Time to one pound: 18 months • pH range: 7 – 8.5	• Not territorial – can be bred with other equal-sized fish. • Good tolerance to water temp variations.	• Require high protein fish food.
4.	Goldfish	• Temperature range: 25 – 37 °C • pH range: 6 – 8	• Very hardy and tolerant of pH changes. • Pretty to look at.	• Not an edible fish.

S.N.	Species	Requirements	Advantages	Disadvantages
5.	Koi or Carp 	• Temperature range: 15 – 25 °C • pH range: 7– 8	• Resistant to most parasites. • Can survive in a wide range of temperatures.	• Not an edible fish • It can produce excess waste as with aging.
6.	Prawn and Shrimps 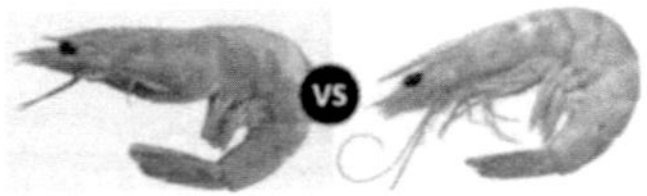	• Temperature range: 14 – 29 °C • Time to one pound: 3 - 6 months • pH range: 6.5– 8	• Minimal maintenance / interference required. • Grow quickly.	• Prawns are susceptible to diseases. • They can attack and eat each other.
7.	Trout 	• Temperature range: 7 – 18 °C • Time to one pound: 4 years • pH range: 6.5– 8	• Good in cooler climates. • Feed on a wide variety of options, including fish, insects, and soft-bodied invertebrates.	• Grows slowly. • Can't be kept with other fish. • Plenty of space required for proper growth.
8.	Crappie 	• Temperature range: 14 – 22 °C • Time to maturity: 2 years • pH range: 6.5– 8.2	• Tolerant of wide temperature changes.	• Need to be very accurate with pH levels. • It cannot be kept together with other fish.

S.N.	Species	Requirements	Advantages	Disadvantages
9.	Guppy	• Temperature range: 23 – 28 °C • Time to one pound: Ornamental fish • pH range: 7– 8	• Good tolerance of pH changes. • Can be combined with other non-aggressive fish.	• Not an edible fish.
10.	Bluegill	• Temperature range: 21 – 23 °C • Time to one pound: 12 months • pH range: 7– 9	• Easy to handle and cope with a wide range of temperatures. • Require little maintenance.	• Need to be fed several times a day to boost growth.
11.	Jade Perch	• Temperature range: 15 – 26 °C • Time to one pound: 12 months • pH range: 6.5 – 8.5	• Very tolerant of temperature changes, pH, and even ammonia levels. • An excellent choice for a beginner.	• Breed fast, which could lead to overstocking if not careful.
12.	Salmon	• Temperature range: 12 – 18 °C • Time to one pound: 2 years • pH range: 7 – 8	• Tolerant of cold conditions.	• High food conversion.
13.	Yellow Perch	• Temperature range: 18 – 21 °C • Time to one pound: 12 months • pH range: 6.5 – 8.5	• Can be grown in shallow water tanks.	• Breeding requires lower temperature about 7.5 °C for a month; this won't be good for plant growth.

6.5 Sustainability of Aquaponics System

In aquaponics farming system, most of the revenue comes from the vegetables which provide a steadier and more predictable revenue stream. On the other hand, fish require almost a year of development before harvest. Hence, growers can start harvesting and selling the vegetables relatively early. After harvesting fish, a new batch of fishing must be started to maintain the steady supply to customer. Aquaponics farm in Philippines conditions, shifting to a higher value fish species like Jade perch along with high demand vegetable makes farming system economical sustainable[2]. The fish component will remain small for an ideal aquaponics farm and have relatively high cost which can only be recovered if high value fish cultured[9]. The main pathways to improve sustainability of aquaponics farming system is optimization of management practices[10]. Fish recirculating aquaculture systems (RAS) wastes in growth media as part of a greenhouse crop system may help to reduce the environmental burden of aquaculture while improving water and nutrient use efficiency in both production systems. Fish effluents suppress fungal diseases and stimulate root growth in tomato and have good potential as an alternative strategy to control soil-borne pathogens in greenhouse tomato[11]. In a home aquaponics system fish to plant ratio (Fish tank volume / Grow bed/area or volume) is normally kept 1:1 but in a commercial setup, it can increase upto 1:3 or 1:4 depending on the fish density and water flow rate. Commercial systems can stock fish as high as 1 fish per 8-10 liters of water.

6.6 Economic Feasibility of Aquaponics System

In many studies, it has been observed that aquaponics system requires higher investment and operating cost. The vegetables production in this system is lower as compared to a hydroponics system. In general, vegetables produced under aquaponics are considered as an organic produce, and sold at 20 per cent premium price, aquaponics becomes profitable[12]. It is important that the grower must understand the basic objective of the aquaponics system. Production of fish alone in such a system is not an economical viable option, but growing crops like lettuce and basil in aquaponics system can be a profitable business. The aquaponics system reveals greater profitability in those areas where fresh produced is very expensive[13,14]. Start-up aquaponics growers may want to explore revenue sources other than just the vegetables and fish produced from aquaponics to enhance economic feasibility[15]. Economic evaluation of three small-scale commercial aquaponics in Hawai'i was analyzed on the basis of (i) profitability, (ii) return on investment and (iii) input requirements. It was found that small-scale commercial aquaponics is reasonably economically feasible for the production of lettuce and tilapia for retail sales. The annual production of lettuce and fish was about 16,247 kg and 1905 kg respectively.

This is equivalent to 89,549 heads of lettuce and 3360 tilapias per year. At each harvesting cycle, 156 kg of lettuce and 37 kg of tilapia can easily be harvested. All three selected aquaponic farms yield profit with an average 10.93 per cent of modified internal rate of returns[16].

6.7 Advantages and Disadvantages of Aquaponics[17]

Aquaponics system is considered as a sustainable production system and is beneficial for the environment. It has some advantages and certain disadvantages, which are as follows:

Advantages

a) The water usage is around 90 per cent lower compared to traditional soil gardening.

b) Growth of plants is significantly faster compared to traditional growing practices using soil media.

c) Aquaponics grown vegetables are bigger and healthier as compared to traditionally grown in soil.

d) There is no need to use chemical fertilizer to feed the plants.

e) There is no need to dispose the fish waste or provide an artificial filtration system.

f) Effective space utilization.

g) Reduced damage from pests and disease.

h) No weeding or bending down on the ground required.

Disadvantages

a) It can be expensive to setup as the system requires pumps, tubing and tanks/beds.

b) Greenhouse is required for a good aquaponics system.

c) Technical knowledge is required for proper monitoring and executing of aquaponics systems.

d) Water quantity and quality is to be constantly monitored to make sure the water quality is fit for fish.

e) Aquaponics requires electric energy input to maintain and recycle water within the system.

f) Root crops can't be grown.

References

1. The Aquaponics Source https://www.theaquaponicsource.com/what-is-aquaponics/ (accessed on July 02, 2020)
2. Bosma RH, Lacambra L, Landstra Y, Perini C, Poulie J, Schwaner MJ, Yin Y. The financial feasibility of producing fish and vegetables through aquaponics. Aquacultural Engineering. 2017; 78:146-54.
3. Blidariu F, Grozea A (2011) Increasing the Economical Efficiency and Sustainability of Indoor Fish Farming by Means of Aquaponics-Review. Animal Science and Biotechnologies 44: 1–8.
4. Bernstein S. Aquaponic Gardening: Growing Fish and Vegetables Together. https://www.motherearthnews.com/organic-gardening/aquaponic-gardening-growing-fish-vegetables-together accessed on July 06, 2020
5. Bradley K. https://www.milkwood.net/2014/01/20/aquaponics-a-brief-history accessed on July 06,2020.
6. Love DC, Fry JP, Genello L, Hill ES, Frederick JA, Li X, Semmens K. An international survey of aquaponics practitioners. PloS one.2014; 9(7): e102662.
7. Mohapatra BC, Chandan NK, Panda SK, Majhi D, Pillai BR. Design and Development of a Portable and Streamlined Nutrient Film Technique (NFT) Aquaponic System. Aquacultural Engineering. 2020;18: 102100.
8. Brooke N. Choosing the Best Fish for Aquaponics. https://www.howtoaquaponic.com/fish/best-fish-for-aquaponics/ accessed on July 06, 2020
9. Engle CR. Economics of Aquaponics. Southern Regional Aquaculture Center (SRAC) 2015; Publication No. 5006.
10. Forchino AA, Lourguioui H, Brigolina D, Pastresa R. Aquaponics and sustainability: the comparison of two different aquaponic techniques using the Life Cycle Assessment (LCA). Aquac. Eng. 2017;77: 80–88.
11. Gravel V, Dorais M, Dey D, Vandenberg G. Fish effluents promote root growth and suppress fungal diseases in tomato transplants. Can. J. Plant Sci. 2015; 95 (2): 427–436.
12. Quagrainie KK, Flores RM, Kim HJ, McClain V. Economic analysis of aquaponics and hydroponics production in the US Midwest. Journal of Applied Aquaculture. 2018; 30(1) :1-4.
13. Bailey DS, Rakocy JE, Cole WM, Shultz KA. Economic analysis of a commercial-scale aquaponic system for the production of tilapia and lettuce. In: Tilapia Aquaculture: Proceedings of the Fourth International Symposium on Tilapia in Aquaculture, Orlando, Florida, pp.1997: 603-612.
14. Holliman JB, Adrian J, Chappell JA. Integration of hydroponic tomato and indoor recirculating aquacultural production systems: an economic analysis. Special Report No. 6, Alabama Agricultural Experiment Station, Auburn, Alabama. 2008
15. Love DC, Fry JP, Li X, Hill ES, Genello L, Semmens K, Thompson R.E. 2015. Commercial aquaponics production and profitability: findings from an international survey. *Aquaculture 2015;* 435: 67-74.
16. Tokunaga K, Tamaru C, Ako H, Leung P. Economics of Small-scale Commercial Aquaponics in Hawai'i. Journal of World Aquaculture Society 2015; 46(1): 20-32.
17. McCarthy M https://sites.google.com/site/aquapanaponics/4-project-updates/ Accessed on July 4, 2020.

7

Secondary Agriculture

7.1 Introduction

Agriculture is the backbone of Indian economy and almost 70 per cent of the total population depends on it. After independence, India achieved self-sufficiency in primary agriculture like food grains, fruits, vegetables and milk, etc. Now it is time to think and focus on secondary agriculture. It is assumed that with the adaption of secondary agriculture, India can achieve the goal of doubling the income of farmers. The secondary agriculture can be defined as, value addition to agricultural products, generating basic facilities for primary processing and add value to the basic agro commodities to allow farmers to get better revenues from their farm produce. It also has job creation opportunity in the rural sector to grow rural economy, which is purely based on agriculture. Further, it can also be elaborated that secondary agriculture as a production activity and invented strategy which includes sustainability of production, monetisation of farmer's produce, straightening of extension service and recognising agriculture as an enterprise. Secondary agriculture can help drive growth of primary agriculture in the three areas i.e. (i) value addition to primary agriculture production, (ii) farmer – industrial linked and (iii) effective utilization of crop residues of primary agriculture.

The value addition to a primary agriculture production can be achieved by establishing direct linkage of farmers with market through grading of agricultural produce, cluster farming, marketing skill and financial literacy. The farmer – industrial linkage is closely associated with rural off-farm activities such as bee keeping, livestock production, etc. it can be promoted as a part of an integrated farming system. The crop residues effectively can be utilized for energy generation. There is scope of establishing rural based agricultural entrepreneurship in the field of biofuel. Agribusiness incubation can help farming community and provide reliable market linkages to improve farmer's realisations.

7.2 Value Addition in Primary Agriculture Production

In the present context, the most prevailing problem faced by Indian farmers is that they are not getting the remunerative price of their farm products. This problem can be resolved át-large extent by processing surplus agricultural produces and marketed inside the country and export to other nations. This is basically a part of value addition of primary product and ensure the nutritional security. The value added agriculture most generally refers to processing of product that increases the value of primary agricultural commodities, and if it coupled with marketing, solves the problem of surplus produces, wastage of produces, and lower price at a large extent. So by and large the value addition of primary agriculture can be defined as "a process in which the value of primary agriculture products can be enhanced by means of processing, packaging, upgrading the quality, increasing shelf life of produces, organic produce and branding to increase the consumer appeals and willingness to pay a premium over similar but undifferentiated products." The agro-processing are set of activities, which are technically viable and economically feasible reform for conservation and handling of farm produce and make it a usable form of food, feed, fibre, fuel and fertilizer[1].

India is one of the largest producers of agricultural products, and post-harvest losses are estimated to about Rs. 926 billion. The post-harvest losses are not only found in perishable commodities like fruits and vegetable but also significantly reported in cereals and pulses. Such losses occur due to lack of proper storage facilities, improper handling of farm products, lack of improper packaging facilities, poor processing facilities, lack of linkage between farmers and processing industries, poor knowledge of post harvesting technologies, poor quality seeds, insufficient food processing technologies, lack of knowledge regarding demand in the market and climate and weather conditions[2]. The Food and Agriculture Organization of the United Nations have estimated that food production will need to grow at globally by 70 per cent to feed world population[3]. In India, crop losses revealed that 4.65 – 5.99 per cent cereals, 6.36 – 8.41 per cent pluses (Gram, tur & others), 3.08 – 9.96 per cent oilseeds, 6.70 – 15.88 per cent fruits, 4.58 – 13 per cent vegetables and 1.18 – 6.51 spices were lost during harvesting, collection, threshing, grading/ sorting, winnowing/cleaning, drying, packaging, transportation and storage depending upon the commodity[4].

7.3 Agro- Processing and Value Addition of Crop and Commodity

7.3.1 Rice processing

Rice is a staple food of more than half of the world's population. Rich in minerals, vitamins and nutrients, it is a good source of the complex carbohydrates. The

United States Department of Agriculture (USDA) estimates that the World Rice Production in 2019/2020 will be 496.08 million metric tons, around 3.23 million tons less than previous month's projection. India accounts for more than 23 per cent of world production, ranking second only to China.

In India, rice is produced in states like Punjab, Haryana, Uttar Pradesh, Bihar, West Bengal, Chhattisgarh, Orissa, Andhra Pradesh, Tamil Nadu and Assam. Total production of rice during 2019-20 is estimated at record 117.47 million tons. It is higher by 9.67 million tons than the five years' average production of 107.80 million tons. Indian basmati rice exports amounted to more than 4.7 billion U.S. dollars in fiscal year 2019. Other rice amounted to about three billion dollars in the same year. With the exception of fiscal year 2017, export value for rice from India had been consistently increasing. Presently, about 1,30,000 rice mills are actively associated in rice processing with capacity in the range[5] of 2 tons / hr to 4 tons / hr.

The efficiency improvement in rice processing mills, significantly reduces the cost of rice production. Considerable amount of energy being consumed during parboiling of paddy rice and drying, which accounting for a large portion of total rice processing costs and its quality. The energy consumed during parboiling process can be met out through rice husk powered gasifier. Addition of such technology in rice processing industry has meaningful improvement and energy economy in rice parboiling mill. Replacing the conventional heat supplying system with rice husk based gasifier shows environmentally friendly and reduce energy outlay and also reduce overall energy cost[6].

7.3.2 Value Addition of Rice

In India, it has been observed that most of the farmers sell their rice to traders after harvesting without any further processing. The cleaning and grading of harvested rice also enhance the market value and will give higher price to the farmers. Further, rice can also be converted into different value added products like puffed rice, flaked rice etc.

7.3.2.1 Brown Rice

It is highly nutritious and non-allergenic in nature. People use brown rice floor because of its various health benefits including anticarcinogenic and hypocholesterolemic effects and have been used as an ingredient to various food products[7]. In the present context, consumers are very much health conscious and looking to the nutritional significance of brown rice and its consumption being encouraged. To increase the nutrient density of rice-based products, various types of products, viz. popped, extruded, bread, cakes, noodles, cookies etc. have been developed from brown rice either alone or in combination with other flours. Value-added products from brown rice have a good potential for consumer acceptance and are regarded as health-promoting functional foods[8].

7.3.2.2 Puffed Rice

It is basically a ready-to-eat snack product obtained by puffing, milled parboiled rice or by sand roasting. Traditionally, rice heated without sand to lower the moisture and then mixed with salt solution and roasted. During roasting rice usually expands and becomes highly porous and crispy. Whereas, in case of paddy made puffed rice, paddy is roasted with sand (Fig. 7.1). The puffed rice is easily digested and has widely accepted by the Indian households. It is an ingredient of Indian popular snack such as bhelpuri chaat. It has good market potential.

Fig. 7.1: Puffed Rice

7.3.2.3 Flaked Rice

Flaked rice is rice, which is parboiled, rolled, flattened, and dried (Fig. 7.2). The flakes of rice swell when added to liquid, whether hot or cold, as they absorb water, milk or any other liquids. The flakes come in different thickness depending on the pressure used in the flattening process. Number of technologies are available for preparation of ready-to-cook lemon bath, curd bath and so on, and it has more export potential.

Fig. 7.2: Flaked Rice

7.4 Oilseed Processing and Value Addition

Edible plant oils and animal fats obtained from milk and milk products are one of the five essential ingredients of human diet, and the others are protein, carbohydrates, minerals and vitamins. In a balanced diet, the oils and fats requirement per person per day is 35 g for vegetarians, 39 g for non-vegetarian and 38 g for average diet.

7.5 Value addition in Floriculture

Floriculture refers to cultivation of flowering and ornamental plants. Flowers provide a good opportunity to convert them into the different valuable products. Floriculture is considered a most profitable agri-business. Floriculture market was worth INR 188.7 Billion in 2019. flowers are integral part of the Indian society, and they are cultivated for social and religious purposes. The demand of cut and loose flowers has significantly increased and it projected that floriculture is one of the important commercial trades in Indian agriculture.

Value addition in floriculture is a process of increasing economic value along with consumer appeal of floricultural commodity. It increases the profit potential of raw commodity when converted into a unique finished product. More profit can be achieved if a producer carefully identifies goods that utilized local resources and that fulfil gap in the market. In the recent year, floriculture has emerged as major diversified business option. The floricultural growers should carefully identify the demands of local market and try to fulfil the gap in the market using local resources. It is essential to establish a supply chain of value added product to extend the spread of the market and give a strong reason to customers to buy the value added products. The product wise business sectors such as cut-flowers, bulbs and tubers, live potted plants, dried plants, dried flower etc. are grouped under this model[9]. In India, it is expanding at a rapid rate and holds good position in Indian market. It may be the most successful component of diversified horticultural industry as it expanded in human lifestyle. The value-added product from floricultural and non-floricultural crops like essential oil of rose, tubes rose, jasmine, marigold etc. flavours, fragnace, insecticidal and nematicidal compound, gulkand, rose water, pigment etc. and plant extract used in medicine and pharmaceutical industry. Mahua flower can also be converted in value added products like muffins, biscuits, cakes and diversification of utilization of mahua from liquor to nutritious food items of mahua flowers. Similarly, lavender is an easy-to-grow, high-value crop that offers a good return with value added products like lavender bags, aromatherapy oil, lavender soap. Some of value added products of flower are listed in Table 7.1. There is good job opportunity in rural areas and women can closely be associated with such activities.

It has been estimated that around 43 per cent of the agricultural labour force of developing countries is directly associated with women. There is a huge potential for development and establishment of women entrepreneurship among rural women. Such entrepreneurship not only increases the national productivity and generate employment but also help to develop economic independence along with personal and social upliftment[10].

Table 7.1: Value Added Products of Some Flowers[11]

Rose	Cut flower, perfume, Rose hips, Rose water, Rose herbal tea, Gulkand, Rose syrup, Rose oil, Rose cream etc.
Jasmine	Jasmin tea, Essential oil, Perfume and incense, Jasmonates, Hair Adornment/ Decoration
Chrysanthemum	Garlands, Potpourri , Edible chrysanthemums, Chrysanthemum insecticides (pyrethrin), Chrysanthemum extracts etc.
Carnation	Carnation concertation and absolutes, Dry flowers, Medicinal carnations, Edible carnations etc.
Anthurium	Standard anthuriums, Obake anthuriums, Tulip anthuriums.
Gladiolus	Bouquets, Flower arrangement, Medicine, Edible gladiolus, Scented gladiolus
Tuberose	Floral ornaments, Essential oils, Medicines, Edible tuberose
Marigold	Phytochemicals, Natural dyes,essential oils,edible product like salad(Tagetes lucida)

7.6 Fish Processing and Value Addition

Fish is a good protein source and only few of food products have comparable nutritional qualities on essential amino acids, essential fatty acids, minerals, vitamins, digestibility, etc. The fisheries and aquaculture production contributes around 1 per cent to India's Gross Domestic Product (GDP) and over 5 per cent to the agricultural GDP. In India, per capita fish consumption in India lies between a range[12] of 5 to 10 kg. The fish production in India increased from 0.75 million tonnes (1950-51) to 12.6 million tonnes (2017-18). At present, India is contributing 6.3 per cent of the global fish production and 5 per cent of the global fish trade[13].

The fish consumption may be increased by accelerating the better use of the existing facility. In the present context, the quality of fish is significantly degraded because of the poor infrastructural facilities of lice ice plant, landing facility etc. Therefore, there is an urgent need to develop good fish handling practices to reduce the wastage of fish, introducing modern preservation technology i.e. freezing, chilling cold storage, improvement in transportation facility with incorporating refrigerated vehicle through better road or train. The hygienic handling and processing techniques and producing different value added products open new employment avenue. Value added fish products may be; i) mince or mince based products ii) battered and breaded or coated products and iii) surimi based products[14]. Through value addition, one can utilize unconventional fish varieties, wide variety of products for those who are afraid about thorns or bone fish, and who don't like fish smell but looking for good nutritive value food. Fish Finger, Fish Cutlet, Fish Nuggets, Fish Pakora, Fish Chakli etc. are some value added product as shown in Fig. 7.3. There is huge scope of value added product based entrepreneurship.

Fish Finger

Fish Cutlet

Fish Nuggets

Fig 7.3: Value Added Product of Fish

7.7 Value Addition in Medicinal and Herbal Plants

India is rich in medicinal and aromatic plants covering a wide area with different climatic conditions. The total world herbal trade is currently assessed at USD 120 billions. India's share in the global export of herbs and herbal products is low due to the following reasons:

(i) Inadequate Agricultural Practices.

(ii) Inadequate Quality Control Procedure.

(iii) Lack of large scale Organic Cultivation.

(iv) Lack of Processing and R&D.

(v) Lack of Standardization in Products, Processes and Services.

(vi) Lack of regulatory framework in trade of Medicinal Plants.

The exports of herbs and value-added extracts of medicinal herbs are gradually increasing over the years. India exported USD 330.18 million worth of herbs during 2017-18 with a growth rate of 14.22 per cent over the previous year. Also, the export of value-added extracts of medicinal herbs / herbal products during 2017-18 stood at USD 456.12 million recording a growth rate of 12.23 per cent over the previous years[15].

Developing nations play major role in supplying the medicinal and aromatic plant material across the world. Hence, developing nations have good opportunity to explore the possibility to established industries on these plants based materials for economic growth and sustainable livelihood of growers.

Study on cultivation and conservation of medical and aromatic plant was conducted in Western Himalaya by Negi et al.[16]. They identified about 152 potential medical and aromatic plant species, which have good medicinal value, high market demand, potential availability and extensive use in a traditional herbal system. Of these, 43 potential medical and aromatic plant were prioritized for cultivation and conservation in Western Himalaya. At high altitude areas of Himalayan regions, return from agricultural production is very low and farmers have limited options to earn money for their daily needs. Collection of medicinal and aromatic plants provides extra, which compensates low agricultural production, and it has been estimated that many village communities derive as much as 10-50 per cent of their household income from the sale of the forest products[17].

Medicinal plants contain chemicals that the medical services use to treat people. Such medicines are derived from plant sap and parts - stems, flowers, leaves, roots - along with synthetic man-made medicines, which are then carefully dosed and packaged into tablets, creams or liquids, injected or applied. Aromatic plants smell nice since 'aroma' means smell or scent. Aromatic plants may also contain medicinal compounds. Medicinal and herbal plants have been widely used for medical purposes not only for humans but for animals as well. Traditional herbal medicines are naturally occurring, plant-derived substances with minimal or no industrial processing that has been used to treat illness within local or regional healing practices. These plants will be continued in various forms (whole, ground, extracts or essential oils) either as therapeutic compounds in pharmaceutical industry or as flavouring, antimicrobial and antioxidant agents in food industry and cosmetics[18].

The essential oils are usually extracted through distillation process, and most of the distillation plants are operated on conventional energy sources. Such distillation plants are a costly affairs to the farmers, and they are unable to extract essential oils of their farm produce at farm. As it is a well-known fact that solar energy is abundantly and freely available and can be used in extract of essential oil through the solar distillation process, establishing the solar distillation at a farm level can enhance the oil extraction, save transportation money, and improve growers income[19]. On-farm solar distillation system was designed and developed by Munir and Hensel (2010) to extract essential oils from medicinal and aromatic plants. A Scheffler fixed focus concentrator was used to achieve desire distillation temperature. They achieve temperature about 300 – 400 °C at focus point with 33.21 per cent system efficiency. Schematic representation of extraction assisted by solar energy is illustrated in Fig. 7.4.

Compound parabolic solar collectors (CPCs) based solar distillery to extract essential oil from two different mint species (Mentha piperita L and Mentha spicata L) was designed and developed by Kulturel and Tarhan[20]. The developed essential oil distillation system was composed of seven compound parabolic solar collector and distillation unit as illustrated in Fig 7.5. The heat transfer oil was heated by CPCs and pumped to a distillation unit. The developed system was in position of produce essential oils in the range of 25 to 40 ml from 5 kg of mint plant per day.

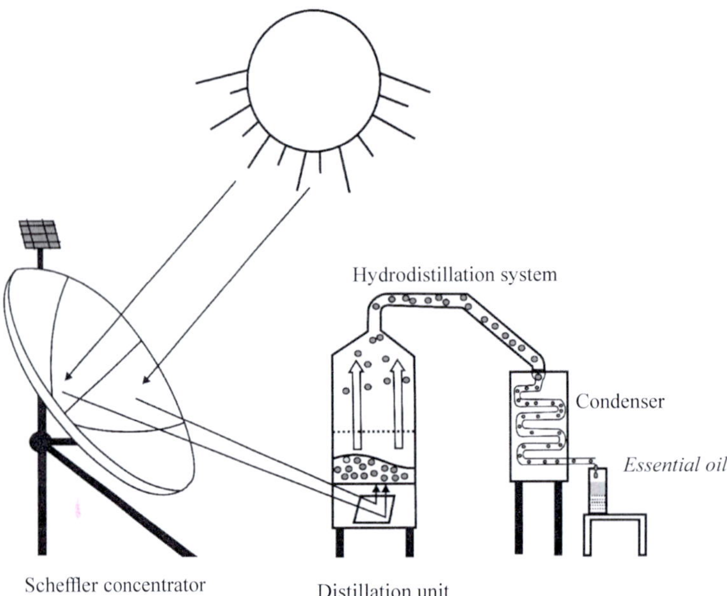

Fig, 7.4: Schematic Representation of Extraction Assisted by Solar Energy

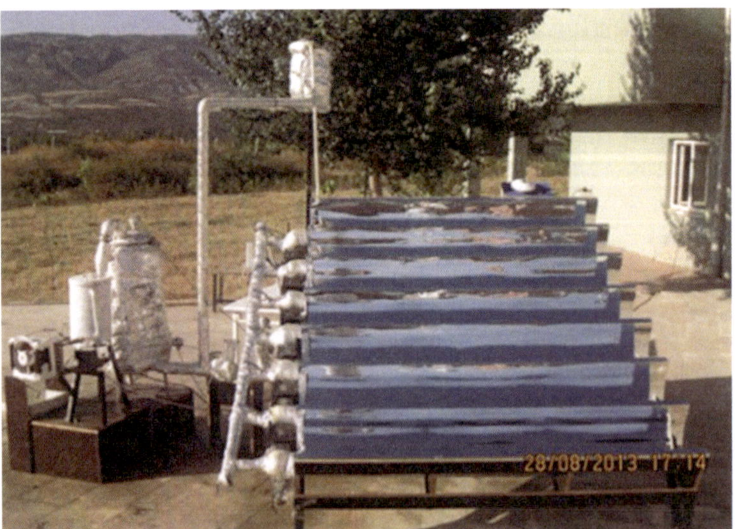

Fig. 7.5: Solar Distillery of Essential Oils with Compound Parabolic Solar Collectors[20]

After extraction of essential oils form medicinal and aromatic plants, considerable volume of residues biomasses is being generated, and these residual biomasses cannot be treated as waste material. It can be further processed and converted into value added products like pellets, biochar etc. Effective utilization and recycling of residual biomass is not only additional economic gain but also provided a practical solution for its proper disposal[21].

7.7.1 Extraction of Essential Oil from Herbal Biomass

The growing emphasis on the demand for green products as natural flavours and fragrance has led to an increase in the importance of herbal and aromatic plants. The consumers today are transitioning to natural extracts from a safety point of view, which has resulted in the increased demand for essential oil in the market[22-24]. The medicinal and aromatic plants constitute a major segment of the biomass resources in India, which is a result of the diverse agro-climatic condition in different regions of the country[25].

Essential oil is one of the products obtained from herbal biomass. They are the hydrophobic secondary metabolites which play a major defensive and attractive role in the interactions between plants and their environment[26,27]. The presence of secondary metabolites in the plants is affected mainly due to the abiotic stresses caused by factors such as temperature, incident light, humidity, nutrients and water and the biotic stresses caused due to damages by living organisms such as bacteria, fungi, virus, parasites and several ecological factors[28]. Essential oils are organic, highly concentrated, volatile, alcohol soluble, mostly colourless, hydrophobic, less viscous than oil, less dense than water, watery texture and highly aromatic and the plants lending essential oils have distinctive fragrance. They are stored and secreted from the tiny secretory structures located in the plant's various parts such as the seeds, roots, barks, stems, leaves, fruits, flowers, zest or seeds[29]. They possess anti-microbial, fungicidal, insecticidal, herbicidal, acaricidal and nematicidal properties as well as therapeutic properties for many pathologies such as anti-inflammatory, anti-allergic, healing, antispasmodic, antiseptic, analgesic, antifungal, antiviral, antimicrobial, strengthen natural immunity and metabolism regulator in humans[30-35].

Lemongrass plant (*Cymbopogon*) is a group of 55 species; it is a widely used indigenous herb in the tropical and semitropical areas, especially in Southeast Asia. Lemongrass essential oils are mainly composed of esters, terpenes, ketones and alcohols[36]. Studies indicate that lemongrass essential oils possess antifungal activity, which is ideal for plant management and disease prevention such as fungal spread; they possess phytochemicals, which have many benefits for human health. They are also found to be safe to be used as an additive in food products[37-39].

It has been estimated that 80 per cent of the total population in the developing countries and 85 per cent of the total population worldwide depend on traditional herbal medicine[22,40]. This impressive demand of essential oil in the local and international market has created tremendous scope for biomass cultivators and enterprises to manufacture herbal products, which calls for developmental research work for the appropriate technology based on the social and economic level of the communities[41,42].

Biomass being one of the most common and widespread renewable resources across the globe, it is a reliable source of energy and gives a better alternative to cope with the environmental challenges faced due to the use of conventional resources. The energy obtained from burning of biomass resources varies from 15.41 to 19.52 mJ/kg[43].

To date, most essential oil extraction plants rely on conventional energy source for their operation. Imbibed with ample biomass resources, they have the potential to meet the energy demands as an alternate energy source. The extraction methods for essential oils will vary depending on the type of herbal biomass, parts of herbal biomass used and accordingly their pre-processing treatment is also determined[44].

7.7.2 Development of Essential Oil Extraction Unit

The essential oil extraction unit to extract essential oil from lemongrass was developed by the Department of Renewable Energy Engineering at the Maharana Pratap University of Agriculture and Technology, Udaipur, Rajasthan, India under the auspices of the All India Coordinated Research Project (AICRP) on Energy in Agriculture and Agro-based industries (EAAI). The assumption made during development is presented in Table 7.2. The schematic of essential oil extraction unit is presented in Fig 7.6.

Table 7.2: Assumptions for Development of Essential Oil Extraction Unit

S No	Particulars	Specifications
1	Product	Lemongrass
2	Specific heat of lemongrass	3.347 kJ/kg K
3	Bulk density of lemongrass	100 kg/m^3
4	Fuel used	Acacia Nilotica
5	Calorific value of fuel	15000 kJ/kg
6	Steam Generator	
	i. Combustion efficiency	30 per cent
	ii. Capacity	20 l
	iii. Diameter of combustion chamber	0.065 m
	iv. Diameter of the boiler	0.235 m

S No	Particulars	Specifications
	v . Temperature of boiler	150°C
	Time of reaction	105 min
7	Charge reactor	
	i. Time of reaction	60 min
	ii. Lemongrass mass	5 kg
	iii. Mass of basket	1 kg
	iv. Diameter of charge reactor	0.30 m
	v. Temperature of charge reactor	100°C
	vi. Asbestos insulation (K)	0.1105 W/mK

Following four main parts which form the base of the essential oil extraction unit:

1. Steam generator
2. Charge reactor
3. Condenser
4. Condensate receiver/ separator

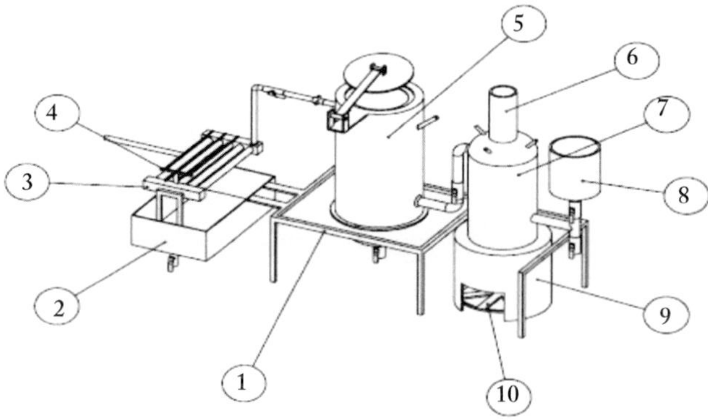

1. Stand **2.** Condenser water tank **3.** Condenser tubes **4.** Shower tube **5.** Charge reactor **6.** Chimney **7.** Steam generator **8.** Water filling tank **9.** Combustion chamber **10.** Grid

Fig 7.6: Schematic Diagram of the System Essential Oil Extraction Unit

Design of Steam Generator

Assume the inner diameter of the combustion chamber as 0.065 m. and the diameter of the boiler as 0.235 m.

Assume the height of the combustion chamber as 0.60 m

Cross-sectional area of the flue gas pipe (A_1)

$A_1 = \pi \times d_1 \times h$

$A_1 = \pi \times 0.065 \times 0.6$

$= 0.1225\ m^2$

Volume of flue gas pipe (V_1)

$$V_1 = \frac{\pi}{4} d^2 \times h$$

$$V_1 = \frac{\pi}{4} \times 0.065^2 \times 0.6$$

$V_1 = 0.00199\ m^3$

Cross-sectional area of the boiler (A_o)

$A_o = \pi \times d_o \times h$

$A_o = \pi \times 0.235 \times 0.6$

$A_o = 0.443\ m^2$

Volume of the boiler(V_0)

$$V_o = \frac{\pi}{4} d^2 \times h$$

$$V_1 = \frac{\pi}{4} \times 0.235^2 \times 0.6$$

$V_1 = 0.026\ m^3$

The required storage capacity water tank is calculated as

$V_w = V_o - V_1$

$V_w = 0.026\text{-}0.00199$

$V_w = 0.024\ m^3$

The schematic diagram of steam generator to supply steam at desired temperature to lemongrass is illustrated in Fig.7.7

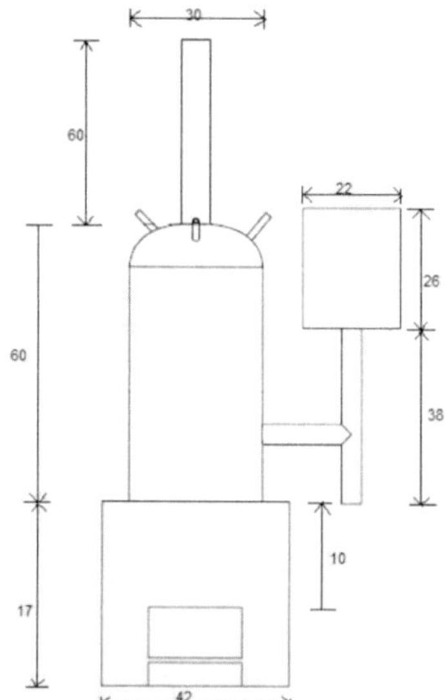

Fig. 7.7: Schematic of Steam Generator

Charge Reactor

The volume of charge reactor (V_3) is calculated by taking the bulk density of lemongrass and desired capacity of charge reactor (kg/m^3).

$$\text{Bulk density} = \frac{\text{Mass}}{\text{Volume}}$$

$$\text{Volume of reactor} = \frac{\text{mass of biomass}}{\text{Bulk density}}$$

$$\text{Volume}(V_3) = \frac{5}{100}$$

$$V_3 = 0.05 \text{ m}^3$$

$$\text{Volume}(V_3) = \frac{\pi}{4} \times d^2 \times h_{reactor}$$

$$h_{reactor} = \frac{0.05 \times 4}{0.3^2 \times \pi}$$

$$h_{reactor} = 0.70 \text{ m}$$

Lid of the charged reactor

Minor axis of the lid (b) =0.33

Major axis of the lid (c) = 0.35

$$A_{lid} = \pi \times \frac{0.33}{2} \times \frac{0.35}{2}$$

$= 0.09\ m^2$

Critical thickness of insulation (r_c)

$$r_c = \frac{\text{Thermal conductivity of asbestos fibre}}{\text{Heat transfer coefficient}}$$

$$r_c = \frac{k}{h_c}$$

$$h_c = \frac{Q_l}{A_{reactor} \times \Delta t}$$

Where,

$Q_l = M_L \times C_p L \times \Delta t$

$Q_l = 5 \times 3.34 \times (100\text{-}20)$

$Q_l = 1336\ kJ$

The surface area of heat transfer is

$A_{reactor} = \pi \times D_{reactor} \times h_{reactor}$

$A_{reactor} = \pi \times 0.38 \times 0.7$

$$h_c = \frac{1336}{\pi \times 0.38 \times 0.7 \times (100 - 20)}$$

$h_c = 19.98\ kJ/m^2\ K$

$h_c = 20\ kJ/m^2\ K$

$h_c = 5.55\ W/m^2\ K$

$$r_c = \frac{0.1105\ W/m\ K}{5.55\ W/m^2K}$$

$r_c = 0.01993\ m$

Outer diameter of charge reactor (D_2)

$D_2 = 2r_c + D_{reactor}$

$D_2 = (2\times0.0119) + 0.3$

$D_2 = 0.3398\ m$

Energy required for the steam generation

$$Q_n = W_t\, C_{pw}\, (T_b - T_a) + W_t\, C_{vw}\, (T_c - T_b) + M_{liq}\, h_l + M_{vap}\, h_g$$

$$= 10 \times 4.15 \times (100 - 20) + 10 \times 1.526 \times (150 - 100) + (7 \times 632.2) + (3\times2745.4)$$

$$=16768.60 \text{ kJ}$$

Fuel required for supply heat

$$\text{Fuel Consumption Rate}\left(\text{FCR}\right) = \frac{Q_n}{\eta_c \times \text{HHV}_{\text{fuel}} \times \text{Time of Reaction}}$$

$$\text{FCR} = \frac{167686}{0.3 \times 15000 \times 105}$$

$$\text{FCR} = 0.0355 \text{ kg/min}$$

The thermal performance of developed system was experimentally evaluated with dried lemongrass having moisture content about 7.80 per cent (Fig. 7.8). The steam temperature at hydrolysis reactor was found 150 °C. About 6.20 kg of sized babool wood was consumed during experiment. The essential oil yield obtained from lemongrass was found in the range of 1.0 to 1.50 per cent.

Fig. 7.8: Developed Essential Oil Extraction Unit

7.8 Utilization of Crop Residues

Agriculture waste management is a big challenge in front of Indian farmers; In India there are 686 MT gross residues available annually from agricultural crop and about 234.5 MT represent the surplus potential. So, proper utilization of surplus crop residue or biomass to produce value added product such as organic biochar which act as a soil amendment and could solve the environmental problem by reducing the green-house gas emission simulteneously helps to farmers getting additional income through more crop production. Crop residues can be converted into biochar through thermo-chemical routes; conversion helps in the managing and handling of biomass. The application of this biochar to soil, improves the physiochemical characteristics of soil because biochar is rich in organic carbon content, which makes the soil more fertile and acts as the carbon sequestration agent over the long term. Furthermore, it reduces the leaching of nitrogen into the ground-water and increases water retention and cation-exchange capacity while moderating the soil's acidity resulting in improve in soil fertility. The farmers are using chemical fertilizers in soil for getting higher production of crop as well as income but overuse of chemical fertilizers can result in hardened soil (due to accumulation of heavy metals in soil), decreased soil fertility, polluted air and water and release of greenhouse gases. There is an urgent need to find an alternative to chemical fertilizers that, ideally, can be sourced in abundant amounts, promotes global food production, enhances capture of CO_2 emissions, and does not affect soil health or damage the environment. Biochar, a pyrogenic black carbon, may play an important role in improving soil health, resulting in higher crop yield and capture of CO_2 emissions. Biochar is produced by heating biomass at high temperatures (300-600°C) in a closed reactor containing no to partial levels of air.

7.8.1 Development of Continuous Biochar Production System

Maharana Pratap University of Agriculture and Technology Udaipur centre of CRP on Energy from Agriculture has developed a continuous biochar production system named as PRATAP BIOCHAR KILN which is capable of the continuous production of biochar from crop residues, with significant recovery of heat (or vapours) and greater flexibility towards biomass feedstocks. The developed kiln has an average operating life of about 10 years and maximum conversion efficiency of 28 to 30 per cent.

7.8.2 Suitability for Specific Condition(s)

The developed PRATAP BIOCHAR KILN was specially designed for continuous production of biochar from different feedstock's having a capacity to process 1-30 kg of biomass per hour. The whole system was made of stainless steel (SS 304) material and required 1 hp electric motor with a gear drive to rotate the screw auger inside the reactor. Groundnut shell, wheat straw, maize straw and woody crop residues may be suitable for the production of biochar in developed kiln. The produced organic carbon rich biochar is more suitable in soil because it acts as a soil amendment and increases the water holding capacity as well as fertility of soil.

7.8.3 Traditional Practice

A very large variety of biomass conversion technologies are being used for the production of biochar in rural areas. There are some batch processes for the production of biochar includes earthen and mound kiln, brick, concrete and metal kiln and retorts. Though the charcoal yield in such a process varies over a low range of 12.5 to 30 per cent, it is still preferred in the countryside because of its low operational and construction cost. The major drawbacks of traditional practices are having lower conversion efficiency and they are exposed to very high level of air pollution simultaneously as it creates adverse effect on environment as well as human being.

7.8.4 Technology Development

A large number of traditional charcoal kilns are available in rural and urban regions but it creates adverse effect on environment. So, to overcome on this problem, MPUAT Udaipur has designed and developed a modern technology which is capable of the continuous production of biochar from crop residues, with significant recovery of heat (or vapours) and greater flexibility towards biomass feedstock. In this reactor, biomass is transported through the heated reactor zone by means of a rotating helical screw. The flow of the biomass through the reactor is adjusted by manipulating the rotational frequency of the screw. The use of mechanical force to move the solid material is advantageous if the particle properties in terms of its dimensions and cohesiveness could cause blockages in gravity fed processes. The developed system has at least 10 years of useful life. The design is simple, use easily available low cost biomass, used SS 304 material for construction and therefore, suit very well for rural, urban as well as industrial applications.

Constructional Details

The schematic of the developed continuous biochar production system is illustrated in Fig.7.9. The required materials used during fabrication are presented in Table 7.3. The technical specifications of the developed continuous biochar production kiln are presented in Table 7.4.

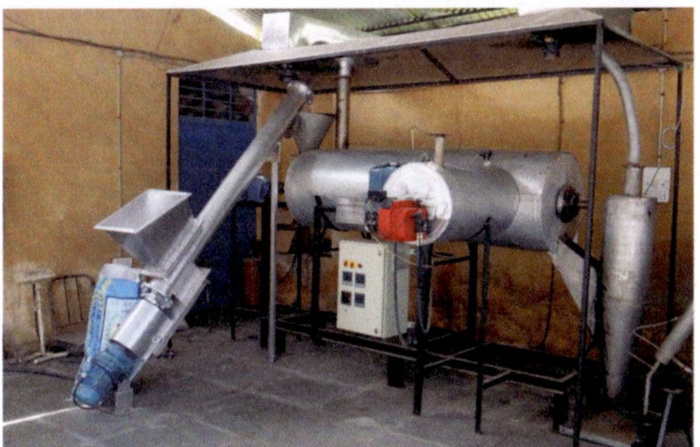

Fig. 7.9: Developed Continuous Biochar Production System

Raw biomass (crop residues) is fed through the hopper which is coupled with auger on the left side of a unit. The main reactor having a length of 2.4 m made of SS 304, inner diameter and thickness are kept as about 0.16 m and 1.5 mm respectively. Inside the main reactor, a screw auger is placed with identical length and material. The diameter of screw auger is kept about 0.15 m. The main reactor receives desired process heat through the outer shell. The outer shell was made of SS 304 with 2.40 m long, and distance between main reactor and outer shell is kept 0.22 m. This is a basically co-centric hollow reactor, where process heat moving is responsible for transforming heat to the main reactor. Cerawool blanket is used as insulation for outer shell to minimize heat losses. K-type thermocouples are placed at different positions to assess the thermal behavior of the developed system.

To rotate a screw auger 1.0 HP electric motor with gear drive is coupled with auger shaft. The biomass loaded in hopper moves to auger and travels throughout the auger length where it converted in biochar and syngas. A cyclone separator is also provided to separate the biochar and syngas. The height of the cyclone separator is kept about 1.51 m, and it is made of SS 304. Material used and the dimensions of the developed continuous biochar production system are presented in Table 7.3 and Table 7.4 respectively.

Table 7.3: Material used for Fabrication of Continuous Biochar Production Unit

Reactor chamber	Material and its thickness
Inner cylinder	SS 304, 1.5 mm
Outer cylinder	SS 304, 1.5 mm
Insulation	Cerawool blanket
Outermost cover over insulation	Aluminium sheet, 0.5 mm
Stand	L shape MS angle
Plenum chamber	
Chamber cylinder	SS 304, 2.0 mm
Insulation	Cerawool blanket
Others	
Screw conveyor	SS 304, 2.0 mm
Hopper	SS 304, 1.5 mm
Cyclone separator	SS 304, 1.5 mm
Outermost cover over insulation	Aluminium sheet, 0.5 mm

Table 7.4: Technical Specifications of Developed Continuous Biochar Production System

Reactor chamber	Values
Diameter of inner cylinder, m	0.16
Diameter of outer cylinder, m	0.38
Distance between inner reactor and outer reactor, m	0.22
Thickness of insulation over the outer cylinder, m	0.04
Plenum chamber	
Diameter of chamber, m	0.42
Insulation over chamber, m	0.04
Length of chamber, m	0.70
Screw conveyor	
Length of screw conveyor, m	2.40
Diameter of screw conveyor, m	0.15
Pitch to pitch distance, m	0.15
Number of pitches	16
Diameter of shaft, m	0.06
Hopper	
Upper diameter of hopper, m	0.50
Bottom diameter of hopper, m	0.15
Height of hopper, m	0.60
Others	
Height of cyclone separator, m	1.51
Volume of screw reactor, m^3/h	0.067
Diameter of outlet gas pipe, m	0.15
Required Diesel burner Specification	
Theoretical Diesel consumption rate, kg/h	0.500
Theoretical power required for Biochar Production Unit, kW	0.708

7.8.5 Characterization of Produced Biochar

The proximate analysis of biomass (Shown in Table 7.5) and characterization of produced biochar (shown in Fig. 7.10) were also calculated. The characterization of produced biochar carried out at SHRIRAM analytical laboratory, New Delhi to detect the composition of minerals. The results found that produced biochar contains 82.90 per cent of carbon, 2.80 per cent hydrogen, 0.80 per cent nitrogen and 9.35 per cent oxygen respectively and the results obtained are mentioned in Table 7.6 and 7.7.

Table 7.5: Proximate analysis of selected biomass sample

S. No.	Biomass sample	Moisture content, %	Volatile matter, %	Ash content, %	Fixed carbon, %	Bulk density, kg/m^3	Calorific value, MJ/kg
1.	Groundnut shells	10.50	75.57	3.52	10.41	80	15.69

Table 7.6: Mineral analysis of produced biochar

Parameters	Carbon (C), %	Hydrogen (H), %	Nitrogen (N), %	Oxygen (O), %	H/C	O/C	Moisture, %
Results	82.90	2.80	0.80	9.35	0.03	0.11	3.82

Table 7.7: Mineral analysis of produced biochar

Parameters	Calcium, %	Iron, %	Zinc, %	Sodium, %	Potassium, %	Sulphate, %	Nitrate, %
Results	0.03	0.0045	0.028	0.09	0.12	0.022	0.0014

If the oxygen to carbon molar ratio is less than 0.2, the half-life is greater than 1000 years. Here, we found the oxygen to carbon ratio as 0.11 so according to the study of Spokas[45] the produced biochar will be more stable in the soil in order to half-life is greater than 1000 years. Similarly, the least hydrogen to carbon molar ratio (H:C) 0.03 was found in produced biochar, which represents the degree of aromatization.

Fig. 7.10: Groundnut Shell and Produced Biochar

Tips for trouble free operation

- Moisture content of crop residue or feedstock should be less than 8 % and well chopped.
- Operating exhaust fans mounted at hood to expel combustion gases from working premise.
- Preheat the inner reactor up to desired operating temperature using diesel burner through plenum chamber.
- Match the RPM screw feeder and auger reactor.
- The outlets of cyclone separator should be cleaned periodically to ensure efficient operation of the kiln otherwise accumulation of gases take place and there is possibility of blast of the whole system.
- Use mask and hand gloves for safety during the operation.

7.9 Pelletization of Crop Residues

With increasing farm mechanisation, large amounts of agricultural crop residues are left in the field to rot, ultimately releasing carbon dioxide to the atmosphere. It could be used to produce pellets, which form fuel for both domestic and industrial furnaces/boiler to generate process heat. The crop residues are to be pre-processed before converting into pellets. The pre-processing focuses on reduction moisture in the range of 8-12 per cent (wb) so the raw material must be put into a dryer to achieve desired moisture level and monitor the particle size. The biomass intended to be used for pellet production should have small particle sizes that is not bigger than[46] 3.2 mm. In case of wheat straw, good quality of pellets can be produced to maintain particle size in the range of 1 to 3 mm and moisture content of about 10 per cent on Wet basis[47]. Flow chart of pellets production through crop residues is presented in Fig. 7.11.

Fig. 7.11: Wood Pellet Lines Flowchart

7.10 Bioethanol from Crop Residues

Bioethanol generation from agricultural crop residues has received considerable attention in current years. Using ethanol as a gasoline fuels additive as well as transportation fuel helps to alleviate global warming and environmental pollution. The agricultural crop residues are renewable, less costly and abundantly available. Surplus crop residues do not have food value and do not require water and energy. Crop residues contain cellulose in a matrix of lignin and hemicellulose. Lignin does not contribute to ethanol generation but cellulose and hemicellulose do[48].

Economic feasibility of bioethanol production is depends on the type of feedstock, type of conversion technology, hydrolysis process and fermentation configuration. By solving bottlenecks of the conversion process, application of novel efficient technology, bioethanol production from agricultural wastes may be successfully developed and optimized in the near future[49]. Flow chart of bioethanol production through crop residues is presented in Fig. 7.12.

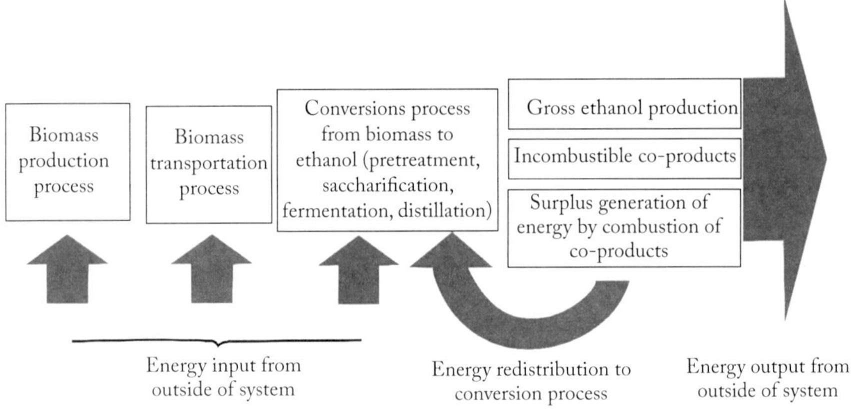

Fig. 7.12: Bioethanol Production from Crop Residues

7.11 Linkage Between Farmers and Industry

India is an agricultural country and agriculture contributes about 16 per cent of GDP. It is the largest sector for employment. Presently in India, agriculture policies fail to recognise how crop choices, input costs, and the supply chain are intertwined, perpetuating marginal farming. There are urgent needs to shift from basic farming to more efficient, sustainable and productive farming. In modern agricultural era, farmers work in the complex web of relationships created by all these individuals and institutions. To make small scale farming more competitive, farmers have started adopting contract farming, where one can access technology, credit marketing channels and information about market price. The contract farming can be defined as agricultural production carried out according to an agreement between a buyer and farmers, which establishes conditions for the production and marketing of a farm product or products. Typically, the farmer agrees to provide agreed quantities of a specific agricultural product. These should meet the quality standards of the purchaser and be supplied at the time determined by the purchaser. In turn, the buyer commits to purchase the product and in some cases, to support production through, for example, the supply of farm inputs, land preparation and the provision of technical advice.

Agriculture and food processing industry are closely linked with the food supply structure and input structure of agricultural products. Agro-based industry itself is an important and dynamic part of the rural economy. The transformation of the agrifood industry, which includes processing, wholesaling, and retailing. An analysis of organizational linkage in the cotton industry in Benin by Sinzogan et al.[50] concluded that neither the reform process, nor the unplanned restructuring is favourable to producers. Further, it was also found that unless farmers are assisted to a change in their production system to release their dependence on production credit and pesticide inputs, they cannot be effective partners. To successfully deal with a range of challenges that confront farmers today, especially the constraints imposed by the small size of holdings of small and marginal farmers, member based Farmer Producer Organizations (FPOs) offer a proven pathway in overcoming these problems.

References

1. Kachru RP (2006) Agro-processing industries in India growth. Status and prospects. Indian Council of Agricultural Research, New Delhi
2. Raut RD, Gardas BB, Kharat M, Narkhede B. Modeling the drivers of post-harvest losses–MCDM approach. Computers and Electronics in Agriculture. 2018; 154: 426-33.
3. Kumari K, Devegoda SR, Pavan MK, Kushwaha S. Post-Harvest Losses in India. Agrobios Newsletter. 2019;18(11): 136-137
4. Lok Sabha, Unstarred Question No. 2460. Department of Agricultural Research & Education Ministry of Agriculture and Farmers Welfare Government of India. Answered on 02/01/2018
5. Singaravadivel K. Rice Processing and Value Addition in India. 2016 http://www.fnbnews.com/Top-News/rice-processing-and-value-addition-in-india-38593 (accessed on May 13, 2020)
6. Beno Wincy W, Edwin M. Techno-economic assessment on the implementation of biomass gasifier in conventional parboiling rice mills. International Journal of Energy Research. 2020; 44(3):1709-23.
7. Wang M, Hettiarachchy NS, Qi M, Burks W, Siebenmorgen T. Preparation and functional properties of rice bran protein isolate. J Agric Food Chem 1999; 47(2):411–416
8. Mir SA, Shah MA, Wani IA. Value-Added Products from Brown Rice. Brown Rice 2017: 203–214.
9. Dawn. Value Addition in Floriculture 2007. https://www.dawn.com/news/247909 (accessed on May 14, 2020)

10. Ranjit Chatterjee R, Das K. Generation of Employment for Women Through Value Addition in Horticultural Crops. Journal of Rural and Community Affairs 2017; II (I): 118-130
11. Mebakerlin, M. S. and Chakravorty, S. Value Addition in Flowers, In: Value Addition of Horticultural Crops: Recent Trends and Future Directions. Edited by Sharangi AB, Datta S. Springer (India) Pvt. Ltd. 2015: 83-99.
12. Press Note. Ministry of Agriculture & Farmers Welfare, Government of India https://pib.gov.in/newsite/PrintRelease.aspx?relid=187305 (accessed on May 14, 2020)
13. Annual Report (2017-18) National Fisheries Development Board, Department of Animal Husbandry & Fisheries, Ministry of Agriculture & Farmers Welfare, Government of India.
14. Datta S. Value added fish products. Retrieved from https://www.researchgate.net/publication/259345025_Value_Added_Fish_Products (accessed May 14 2020).
15. Press Note "Export of Herbs and Herbal Products" Ministry of Commerce & Industry Government of India (https://pib.gov.in/Pressreleaseshare.aspx?PRID=1558955 (accessed on May 18, 2020)
16. Negi VS, Kewlani P, Pathak R, Bhatt D, Bhatt ID, Rawal RS, Sundriyal RC, Nandi SK. Criteria and indicators for promoting cultivation and conservation of Medicinal and Aromatic Plants in Western Himalaya, India. Ecological indicators. 2018; 93: 434-46.
17. Chauhana RS, Nautiyala BP, Nautiyala MC. Trade of threatened Himalayan medicinal and aromatic Plants-socioeconomy, management and conservation issues in Garhwal Himalaya, India. Global Journal of Medical Research. 2013; 13(2): 9-18.
18. Giannenas I, Sidiropoulou E, Bonos E, Christaki E, Florou-Paneri P. The history of herbs, medicinal and aromatic plants, and their extracts: Past, current situation and future perspectives. InFeed Additives 2020: 1-18. Academic Press.
19. Munir A, Hensel O, Scheffler W, Hoedt H, Amjad W, Ghafoor A. Design, development and experimental results of a solar distillery for the essential oils extraction from medicinal and aromatic plants. Solar energy. 2014; 108: 548-59.
20. Kulturel Y, Tarhan S. Performance of a Solar Distillery of Essential Oils with Compound Paraboilc Solar Collectors. Journal of Scientific & Industrial Research 2016; 75: 691 - 696
21. Saha A, Basak BB. Scope of value addition and utilization of residual biomass from medicinal and aromatic plants. Industrial Crops and Products. 2020; 145:111979.
22. Verma S, Singh SP. Current and future status of herbal medicines. Veterinary world. 2008; 1(11): 347.
23. Rao, BRR, Rajput DK, Nagaraju G, Adinarayana G. Scope and Potential of Medicinal and Aromatic Plants Products for Small and Medium Enterprises. Journal of Pharmacognosy 2012; 3 (2): 112–114.
24. MH SK, Srinivas MP. Pests attacking medicinal and aromatic plants in India: A review. Journal of Entomology and Zoology Studie. 2018; 6(5): 201-5.
25. Deshpande RS, Neelakanta NT, Hegde N. Cultivation of medicinal crops and aromatic crops as a means of diversification in agriculture. Agricultural Development and Rural Transformation Unit, Institute for Social and Economic Change 2006
26. Figueiredo AC, Barroso JG, Pedro LG, Scheffer JJ. Factors affecting secondary metabolite production in plants: volatile components and essential oils. Flavour and Fragrance journal. 2008; 23(4): 213-26.
27. Wink M. Modes of action of herbal medicines and plant secondary metabolites. Medicines. 2015; 2(3): 251-86.
28. Duarte MC, Duarte RMT, Rodrigues RAF, Rodrigues MVN. Essential Oils and Their Characteristics. In Essential Oils in Food Processing: Chemistry, Safety and Applications, edited by Seyed Mohammed Bagher Hashemi, Amin Mousavi Khaneghah, and Anderson de Souza Sant'Ana, 2017:1–19. Wiley-Blackwell.
29. Djilani A, Dicko A. The therapeutic benefits of essential oils. Nutrition, well-being and health. 2012; 7:155-79.

30. Hammer KA, Carson CF, Riley TV. Antimicrobial activity of essential oils and other plant extracts. Journal of applied microbiology. 1999; 86(6): 985-90.
31. Güllüce M, Sökmen M, Daferera D, Ağar G, Özkan H, Kartal NU, Polissiou M, Sökmen A, Şahin F. In vitro antibacterial, antifungal, and antioxidant activities of the essential oil and methanol extracts of herbal parts and callus cultures of Satureja hortensis L. Journal of agricultural and food chemistry. 2003; 51(14):3958-65.
32. Mimica-Dukic N, Bozin B, Sokovic M, Simin N. Antimicrobial and antioxidant activities of Melissa officinalis L.(Lamiaceae) essential oil. Journal of agricultural and food chemistry 2004; 52(9):2485-9.
33. Tanu B, Harpreet K. Benefits of essential oil. Journal of Chemical and Pharmaceutical Research. 2016; 8(6): 143-9.
34. Fatima T, Beenish BN, Gani G, Qadri T, Bhat TA. Antioxidant potential and health benefits of cumin. Journal of Medicinal Plants. 2018; 6(2): 232-6.
35. Han X, Eggett DL, Parker TL. Evaluation of the Health Benefits of a Multivitamin, Multimineral, Herbal, Essential Oil–Infused Supplement: A Pilot Trial. Journal of dietary supplements. 2018; 15(2): 153-60.
36. Shah G, Shri R, Panchal V, Sharma N, Singh B, Mann AS. Scientific basis for the therapeutic use of Cymbopogon citratus, stapf (Lemon grass). Journal of advanced pharmaceutical technology & research. 2011; 2(1):3.
37. Tzortzakis NG, Economakis CD. Antifungal activity of lemongrass (Cympopogon citratus L.) essential oil against key postharvest pathogens. Innovative Food Science & Emerging Technologies. 2007; 8(2): 253-8.
38. Hanaa AM, Sallam YI, El-Leithy AS, Aly SE. Lemongrass (Cymbopogon citratus) essential oil as affected by drying methods. Annals of Agricultural Sciences. 2012; 57(2): 113-6.
39. Balakrishnan B, Paramasivam S, Arulkumar A. Evaluation of the lemongrass plant (Cymbopogon citratus) extracted in different solvents for antioxidant and antibacterial activity against human pathogens. Asian Pacific Journal of Tropical Disease. 2014; 4(1): S134-9.
40. Kumari R, Kotecha M. A review on the standardization of herbal medicines. Int J Pharm Sci Res. 2016; 7(2): 97-106.
41. Do TK, Hadji-Minaglou F, Antoniotti S, Fernandez X. Authenticity of essential oils. TrAC Trends in Analytical Chemistry. 2015; 66: 146-57.
42. Sanganeria, S. 2005. "Vibrant India." Perfumer and Flavorist 30 (7): 24–35.
43. Erol M, Haykiri-Acma H, Küçükbayrak S. Calorific value estimation of biomass from their proximate analyses data. Renewable energy. 2010;35(1):170-3.
44. Elmoubarki R, Taoufik M, Moufti A, Tounsadi H, Mahjoubi FZ, Bouabi Y, Qourzal S, Abdennouri M, Barka N. Box-Behnken experimental design for the optimization of methylene blue adsorption onto Aleppo pine cones. J. Mater. Environ. Sci. 2017; 8(6):2184-91.
45. Spokas KA (2010) Review of the stability of biochar in soils: predictability of O:C molar ratios. Carbon Manag 1(2):289–303
46. Mani S, Tabil LG, Sokhansanj S. An overview of compaction of biomass grinds. Powder Hand. Process. 2003; 15(3): 160–168.
47. Stelte W, Holm JK, Sanadi AR, Barsberg S, Ahrenfeldt J, Henriksen UB. A study of bonding and failure mechanisms in fuel pellets from different biomass resources. Biomass Bioenergy 2011; 35(2): 910–918.
48. Reijnders L. Ethanol production from crop residues and soil organic carbon. Resources, conservation and recycling. 2008; 52(4): 653-8.
49. Sarkar N, Ghosh SK, Bannerjee S, Aikat K. Bioethanol production from agricultural wastes: an overview. Renewable energy. 2012; 37(1): 19-27.

50. Sinzogan AA, Jiggins J, Vodouhe S, Kossou D, Totin E, Van Huis A. An analysis of the organizational linkages in the cotton industry in Benin. International Journal of Agricultural Sustainability. 2007; 5(2-3): 213-31.

8

Phytotronics

8.1 Introduction

In the modern world, the most elementary desire of every common man is to have sufficient food and mental peace. With the increasing world population, pressure on both peace and availability of food are continuously increasing. Advancing agricultural research can explore a possibility to feed to all living beings. Agricultural crops are highly susceptible to climatic conditions and crop productivity which are mainly influenced by extreme weather conditions; including temperature and water availability. In open field conditions, during long dry weather condition, soil moisture reduces and increases the transpiration[1].

Plants are very important to all forms of life on the earth because they consume carbon dioxide from the atmosphere and produce oxygen. In addition, plants make up the base of the food web. Basically, plants need five things in order to grow and these are: sunlight, proper temperature, moisture, air and nutrients. These five things should be provided by the natural or artificial environments where the plants live. If any of these elements are missing, they can limit plant growth. If any environmental factor is less than ideal, it limits the plant›s growth. People began putting plants indoor to avoid the damage from cold weather and this practice led to development of greenhouse. With the improvement in living standard and upgradation in agricultural technologies, off seasoned cultivation is also possible by providing desired artificial environmental conditions. The environmental stress weakens a plant and makes it more susceptible to disease or insect attack. In the modern engineering era, plant production systems have become more sophisticated. The desired climatic condition to grow plants and its control have changed from manual to automated digital operations and control computers have become faster and more capable[2].

Environmental control techniques are essential for protected cultivation and are not only for greenhouse's cultivation but are also applicable to tissue culture and space farming as well. Environmental control in the present protected crop cultivation practices has been profusely supported by computer and sensor developments which is sometimes aslo called controlled environment agriculture[3].

Controlled Environment Agriculture (CEA) is basically a technology-based approach toward food production developed with the aim to provide protection from pest or disease to maintain optimal growing conditions throughout the growth of the crop and increase yield or save costs. Production takes place within an enclosed growing structure such as a greenhouse or building. The CEA also protect plant from air pollution. It is a well known fact that air pollutants have a negative impact on overall plant growth, primarily through interfering with resource accumulation. Since leaves are in close contact with the atmosphere, many air pollutants, such as O_3 and NOx, affect the metabolic function of the leaves and interfere with net carbon fixation by the plant canopy[4].

8.2 The Concept of a Phytotron

Phytotron basically is an enclosed research greenhouse extensively used for investigating the interactions between plants and environment and to know how various environmental factors affect plant growth and development. Phytotrons are growth chambers that provide artificial controlled environmental conditions for growing plants. In other words, it can be defined as a facility where a number of Growth Chambers and Greenhouses are organized in such a way that different environmental factors can be simulated for research studies simultaneously.

The first phytotron was built under the direction of Frits Warmolt Went at the California Institute of Technology in 1949. It was funded by the Harry B. Earhart Foundation and was officially known as the Earhart Plant Research Laboratory. Phytotron spread around the world between 1945 and the present day to Australia, France, Hungary, The Soviet Union, England and the United States. Moreover, they have spurred variants such as the Climatron at the Missouri Botanical Garden, the Biotron at the University of Wisconsin-Madison, the Ecotron at Imperial College London and the Brisatron at the Savannah River Ecology Laboratory.

The phytotron in Stockholm offered a humidity controlled room and a custom built computer as well as a low temperature room that extended the temperature range down to -25°C for the study of Nordic forests. After that, phytotron technology compressed whole environments into smaller cabinets able to be set to any desired combination of environmental conditions.

National Phytotron Facility in India was established in August 1997 at Indian Institute of Agricultural Research Institute, New Delhi. This is the first facility of its kind in the country to study the live responses of plants under controlled conditions and the possible impact of climate change and greenhouse gases. The facilities were developed to support Indian Agriculture in improving the crop productivity through better understanding of plant growth under precisely controlled environmental conditions, aimed at the evolution of new varieties and technologies. The phytotron facility is accessible to Indian Council of Agricultural Research Institutions, all States Agricultural Universities, Agriculture related plant science researchers from National and International Universities and Institutes, and Private industries involved in Agri-business on the collaborative basis. Phytotrons are powerful tools to analyse day length and desirable temperature for reproduction of plants and provide opportunities for modification and acceleration of crop life cycles in plant breeding programs[5].

8.3 Plant Growth Chamber

Plant growth chamber is basically an environmental test chamber which is specifically designed to create desirable environmental conditions responsible for effective seed germination and plant growth. In such plant growth chambers; temperature, relative humidity, lights air flows and carbon dioxide concentration is monitored and maintained in such a way that required essential environmental conditions can be created for examining growth of a particular plant as shown in Fig. 8.1. Plant growth chambers are commonly used in the agricultural fields in order to conduct research on crop productivity. Plant related agriculture research and development is the prime application of these units. Plant breeding, plant nutrition, photosynthesis are some of the areas where plant growth chambers play an important role. Plant categories may vary such as cereal, arabidopsis, algae, tobacco, physcomitrella, fungi, medicago, weeds, soybeans, potatoes, other commercial crops, horticulture plants, etc.

It is a well-known fact that plant growth chambers are decisive in generating reproducible observations in plant biology research. Commercially available plant growth chambers are expensive and require high operating cost to maintain the desired environment. To overcome such limitation, Katagiri et al.[7] designed and constructed an inexpensive plant growth chamber with material cost of USD 2,300 with commercial plant growth could be USD 40,00 or more. The designed plant growth chamber has a total growth area of 4.5 m^2 with 40 cm high clearance. Further, they are able to controlled temperature and maintain relative humidity at about 22.2°C±0.8°C, and 75%±7% respectively.

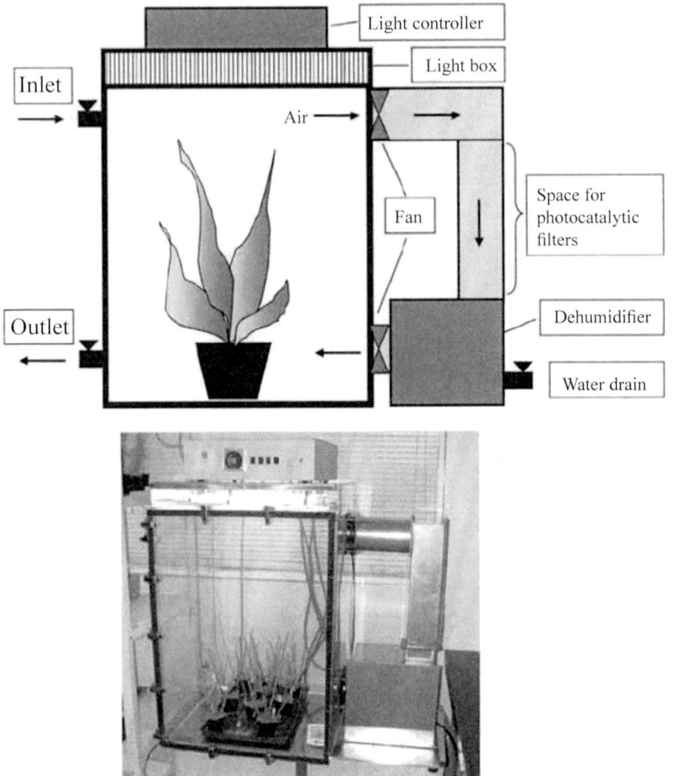

Fig. 8.1: Schematic of Plant Growth Chamber[6]

8.4 Lighting in Plant Growth Chamber

Light-Emitting Diodes (LEDs) are the most suitable energy-efficient light source for plant growth chambers. It directs all their light towards the plant without the reflectors required by other light sources that emit radially. LEDs emit considerably more light per unit of input energy than other lighting sources and produce less radiant. Therefore, can be placed closer to plant material. With less heat generation, the cooling requirement for the controlled environment is reduced and the total energy used by an LED-equipped chamber is substantially reduced. Red or far-red LED light sources are capable in modulating the phytochrome response during de-etiolation, root growth and greening and senescence (Fig. 8.2). LED lighting systems are well suited for plant growth chambers as they have an extraordinary life (about 100,000 hours), require little maintenance and use negligible energy. These factors render LED-based light strategies, particularly appropriate for space-biology as well as terrestrial applications. Multipurpose and inexpensive LED array platform is required for individual wavebands to produce a series of light combinations consisting of various quantities and qualities of individual wavelengths[8]. These lights are controlled through a cyclic timer for 24X7 a week and can be easily changed or replaced at the user level.

Fig. 8.2: LEDs Lighting System for Plant Growth Chamber

Monostori et al.[9] investigated the effects of LED on growth and development, leaf photosynthesis, thiol and amino acid metabolism as well as grain yield and flour quality of wheat. They reported that photosynthetic activity, the number of tillers, biomass and yield significantly increased with increas in the LED light intensities. Optimum photosynthetic activity was recorded when at least 50 per cent of red light was applied during cultivation. A high proportion of blue light extended the juvenile phase, while the shortest flowering time was observed when the blue to red ratio was around one. Further, it was observed that the Blue and Pink regimens of LEDs had the greatest effect on flour quality. This may be due to decay of starch content in the flour probably due to lower CO_2 assimilation capacity of leaves in plants grown under Blue regimen.

8.5 Temperature in Plant Growth Chamber

Temperature is one of the important factors which affects the overall plant growth. These include total plant growth, a degree of deviation from normal' growth, breaking of bud dormancy and flowering. It directly controls the biochemical reaction rates in the various metabolic processes and indirectly through the development of water stress caused by the physical process of transpiration. The measurement of temperature in plant growth temperature largely depends on the sensitivity and response time of selected sensor. The plant leaf is a good conductor of heat and specific heat is high due to water content. Therefore, change in leaf temperature is observed much slower than in the air around it. Further, temperature controls plant growth and development by modulating the rate of the numerous physical and biochemical processes that are a part of the physiology of a plant. Thermocouples or thermistors type temperature sensors are usually used to check the temperature of a plant growth chamber. Resistance type temperature indicators may be ideal but in controlled environment room practice, they have proven troublesome. Iron-constantan and copper-constantan thermocouples are the most commonly used, but under some conditions the iron has a tendency to corrode, and high thermal conductivity of copper may be undesirable under certain circumstances. Overall Chromel-alumel thermocouple would then be a better choice[10].

The temperature control system mainly consists of temperature controller and temperature sensor. Normally, in plant growth chamber, temperature is kept in the range of 5°C to 60°C and controlled through imported PID controller which features excellent accuracy and uniformity with PT100 temperature sensors. Tubular air heaters are used for maintaining growth chamber temperature as shown in Fig 8.3.

Fig. 8.3: Tubular Air Heater and Temperature Indicator

8.6 Relative Humidity and Controlled Environments

Relative humidity (RH) is the ratio of the partial pressure of water vapour to the equilibrium vapour pressure of water at a given temperature. It depends on temperature and the pressure of the system of interest. The same amount of water vapour results in higher relative humidity in cool air than warm air. Relative humidity plays major role in overall plant growth. If the conditions are too humid, it may promote the growth of mold and bacteria that cause plants to die and crops to fail, as well as conditions like root or crown rot. Humid conditions also invite the presence of pests such as fungus gnats, whose larva feed on plant roots and thrive in moist soil.

Relative humidity has great impact on the environment's ability to control temperature, life and growth of its contents. Relative humidity and temperature are closely tied in the small volume of a growth chamber as even small temperature changes cause significant changes in the relative humidity level.

There are two important components which controll or maintain the relative humidity levels in an incubator or growth chamber. The first component add humidity to the environment either through a Pan-type or reservoir humidifier or Ultrasonic humidifier or Spray nozzle / spray mist. The second component which controll relative humidity is dehumidification. In dehumidification process the moisture or water vapour or the humidity is removed from the air keeping its dry bulb (DB) temperature constant. This system comprises of an electric heater and one dehumidifying evaporator. As and when dehumidification is required, the temperature of the evaporator will drops to the point where the chamber moisture condenses on the coil. The heater acts to reheat the dried air back to the programmed temperature before the air is returned to the chamber environment[11].

A desiccant dehumidifier can be coupled with the plant growth chamber for the purpose of dehumidification. It is basically consisting of a corrugated wheel usually made of ceramic material, so that air can overtake along the corrugations. Silica gel a desiccant material is impregnated into the wheel. As the air passes through the wheel, moisture is attracted from the air onto the desiccant. The wheel rotates slowly between two air streams: process air stream and reactivation air stream as shown in Fig. 8.4. Process air is the ambient air dried by the desiccant before being introduced to the component or system to be protected. During this process, as a result of dehumidification, process air gets warmer. Reactivation air is heated and runs through the wheel in the opposite direction of the process air. Its role is to transfer heat to the wheel, heating the desiccant to remove and carry away its moisture so that the desiccant can be reused to collect more moisture from the process air and the cycle be repeated[12].

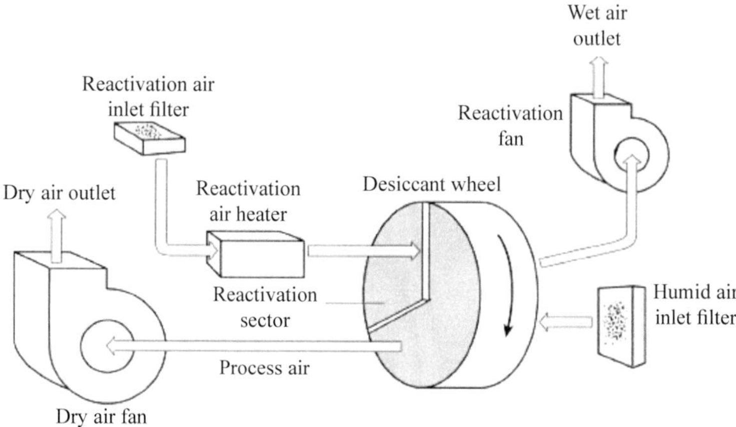

Fig. 8.4: Schematic of a desiccant dehumidifier system

Controlling of relative humidity in a plant growth chamber above 50 % is comparatively a much easier task. Therefore, humidity control in most of the controlled-environment rooms of phytotron covers the range of 50 to 90 %, with a few chambers that can provide 100 % and a few in which humidity may be maintained as low as 15 %.

8.7 Air flow in Plant Growth Chamber

Many researchers and scientist have recognized the importance of air movement in controlling the plant growth in field conditions. The importance of measuring and controlling air movement in the plant growth chamber is being investigated from last 20 years. Plant growth chambers are commercially available with downward air flow and upward airflow. In downward airflow, plant growth chambers come through air diffusers at the top of the chamber's walls. Air-handling plenums provide a stable and uniform plant canopy temperature. Downward airflow plant growth chambers are ideal for growing plants in trays and pots. Upward airflow operated in conjunction with tapered air-handling plenums through the chamber floor as shown in Fig. 8.5. A constant combination of air volumes and velocities ensures a consistent environment. Plant growth chambers that use upward airflow are best for growing plants in pots or doing tissue culture work. The biological effects of air movement in a plant growth chamber are significantly lower than field conditions. Air movement influences plant growth by determining the extent of: (a) sensible or conductive heat transfer from plant surfaces, substrate and plant containers (b) transpiration, evaporation and latent heat transfer from plant surface and substrate. The average air flow rate in the plant growth chamber should be measured at a minimum of five different positions uniformly spaced throughout the plant growing surface[13]. In a plant growth chamber, a forced convection based ventilation and circulation system is usually used to control the growing environment and maintain climate uniformity[14].

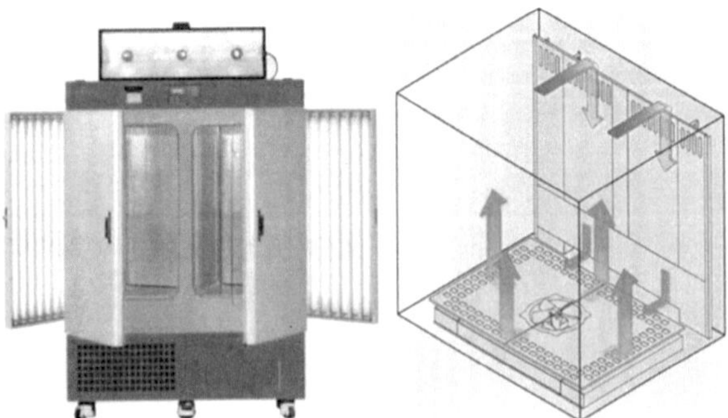

Fig. 8.5: Efficient air flow in Plant Growth Chamber

8.8 Carbon Dioxide Enrichment

Rapid industrialisation, population growth and burning of fossil fuels are widely evident that global atmospheric carbon dioxide (CO_2) concentration has dramatically increased. This high concentration of CO_2 in atmosphere may not only cause climate warming, but also cause deep impacts on the net primary productivity of agricultural and natural ecosystems[15].

The plant growth is significantly enhanced by enriching the air with carbon dioxide. The plant growth stimulation was mostly obtained by sowing crop species in controlled environment, especially greenhouse to determine how to increase yield with CO_2 fertilization. Phytotron is an integrated collections of controlled growth facilities. The major advantage of phytotron is that it has multiple chamber or growth room may be used to create matrices of environmental variables[16].

The concentration of CO_2 in growth chambers filled with plants often is reduced far below normal. Most plant growth chambers are well equipped with some means of introducing air from outside to prevent the CO_2 concentration from falling to excessively low levels by the addition of CO_2 from a compressed gas tank[17].

Atmospheric carbon dioxide level in air is about 340 ppm by volume. Plants grow well at this level but as CO_2 levels are raised by 1,000 ppm, photosynthesis increases proportionately resulting in more sugars and carbohydrates available for plant growth. If the CO_2 levels decline until the threshold of 150 ppm below which plants begin first to starve, subsequently stop growing altogether, and then die.

Studies have shown that higher concentrations of atmospheric carbon dioxide affect crops in two important ways: they boost crop yields by increasing the rate of photosynthesis, which spurs growth and they reduce the amount of water crops lose through transpiration. Plants transpire through their leaves which contain tiny pores called stomata that open and collect carbon dioxide molecules for photosynthesis. During that process, they release water vapour. As carbon dioxide concentrations increase, the pores don't open as wide, resulting in lower levels of transpiration by plants and thus increased water-use efficiency[18].

8.9 Advantages of a Growth Chamber

a) Higher yields due to accelerated cultivation process.

b) Year-round production is quite possible.

c) Better and more reliable crop quality.

d) Effective use of space.

e) Multiple layers cultivation.

f) Efficient use of energy

g) No light pollution.

h) Effective utilization of surplus heat for sustainable production.

References

1. Nemeskéri E, Remenyik J, Fári M. Studies on the Drought and Heat Stress Response of Green Bean (*Phaseolus Vulgaris* L.) Varieties under Phytotronic Conditions. Acta Agronomica Hungarica 2008; 56(3): 321–328.
2. Albright LD, Arvanitis KG, Drysdale AE. Environmental control for plants on earth and in space. IEEE Control Systems Magazine. 2001; 21(5): 28-47.
3. Takakura T. Greenhouse production in engineering aspects. IFAC Proceedings Volumes. 1991 Sep 1;24(11):19-21.
4. Weber J, Tingey D, Andersen C. Plant Response to Air Pollution. U.S. Environmental Protection Agency, Washington, D.C.,2002: EPA/600/A-93/050 (NTIS PB93167260).
5. Evans LT. The Role of Phytotrons in Agricultural Research. International Rice Research Institute. Climate and Rice. Los Banos, Philippines. 1976
6. Kikuchi J, Hirayama T. Practical aspects of uniform stable isotope labeling of higher plants for heteronuclear NMR-based metabolomics. InMetabolomics 2007: 273-286. Humana Press.
7. Katagiri F, Canelon-Suarez D, Griffin K, Petersen J, Meyer RK, Siegle M, et al. Design and Construction of an Inexpensive Homemade Plant Growth Chamber. PLoS ONE 2015;10(5): e0126826.
8. Folta KM, Koss LL, McMorrow R, Kim HH, Kenitz JD, Wheeler R, Sager JC. Design and fabrication of adjustable red-green-blue LED light arrays for plant research. BMC plant biology. 2005; 5(1):17.
9. Monostori I, Heilmann M, Kocsy G, Rakszegi M, Ahres M, Altenbach SB, Szalai G, Pál M, Toldi D, Simon-Sarkadi L, Harnos N. LED lighting–modification of growth, metabolism, yield and flour composition in wheat by spectral quality and intensity. Frontiers in plant science. 2018; 9: 605.
10. Downs RJ, Hellmers H. Environment and the Experimental control of plant growth. Academic Press Inc. (London) Ltd 1975
11. Percival Scientific Humidity in Growth Chambers. https://www.percival-scientific.com/blog/humidity-in-growth-chambers/ accessed on June 8, 2020
12. Desiccant dehumidifiers. https://www.enggcyclopedia.com/2012/02/desiccant-dehumidifiers-work/ accessed on June 8, 2020
13. Downs RJ, Krizek DT. Air movement. In R. W. Langhans, & T. W. Tibbitts (Eds.), Plant growth chamber handbook 1997: 87-104. No. 340, Iowa Agr. & Home Econ. Expt. Sta. Spec. Rpt. No. 99, Ames, IA: North Central Regional Res. Publ.
14. Zhang Y, Kacira M, An L. A CFD study on improving air flow uniformity in indoor plant factory system. Biosystems Engineering. 2016; 147: 193-205.
15. Zheng Y, Li F, Hao L, Shedayi AA, Guo L, Ma C, Huang B, Xu M. The optimal CO_2 concentrations for the growth of three perennial grass species. BMC plant biology. 2018 ;18(1): 27.
16. Drake BG, Rogers HH, Allen Jr LH. Methods of exposing plants to elevated carbon dioxide. Direct effects of increasing carbon dioxide on vegetation. 1985: 238:11.

17. Patterson DT, Hite JL. A CO2 monitoring and control system for plant growth chambers. Ohio J. Sci.1975; 75(4): 190.
18. NASA Study: Rising Carbon Dioxide Levels Will Help and Hurt Crops. https://www.nasa.gov/feature/goddard/2016/nasa-study-rising-carbon-dioxide-levels-will-help-and-hurt-crops accessed on June 10, 2020

9

Artificial Intelligence in Agriculture

9.1 Introduction

The increase in world's population and respective food demand has brought the global concern. The climate change alone has put a tremendous burden on the farming communities, and has become the major challenge to feed the burgeoning population[1]. It has been estimated that by the end of the year 2050, the global population will be around 10 billion[2], meanwhile, the food demand will also be increased by 70-100 %[3,4]. The traditional method of agricultural practices are not sufficient enough to meet the rising food demand. Therefore, to meet the increasing food demand, agriculture is intensified by the use of chemical fertilizer, pesticides, herbicides, which deteriorat the soil health[5]. Thus, new automated/ mechanised systems have been introduced into the agriculture sector to ensure precision farming. These systems are quick in action, and satisfing the requirements along with generation of billions of employment worldwide[2]. In a developing country like India, 65-70 per cent of the total population resides in rural areas[6], of which, most of the households are dependent on agriculture. The agriculture sector in India accounts for 18 per cent of the country's GDP, and provide employment to 50 per cent of the total work force. The agriculture sector is one of the most sensitive sectors of Indian economy which supports all other sectors, and spreading its benefit to the far-reaching areas. Therefore, it becomes important for the developing country like India to develop agriculture sector to boost-up the rural economy[5,7,8].

With the advent of technology, a rapid transformation has been occurred in various industries, while agriculture remains the least digitized[2]. Though, the application of aircraft/ drones in agriculture for fertilizer/ pesticides applications, image capturing, access to real-time quality data has adopted in some developed countries like Australia, USA, Japan, and some of the European countries. Moreover, these unmanned aerial vehicles could provide data of soil health, monitor crop health, assist in irrigation scheduling, yield estimation, and weather data analysis too[1]. However, such development in agriculture is still lacking in developing countries. The agriculture development in the modern era could be done by taking the biological information and agricultural knowledge through integration of modern agriculture information technology with traditional agricultural practices. These knowledge integrations with mobile internet, cloud computing, and machine learning could enable to control the agricultural production system over stressed climate condition easily. Thus, Artificial Intelligence (AI), which is a sort of machine intelligence, an inbuilt set of programme and algorithms similar to the human brain in the current world is gaining more attention in agricultural application for precision farming[9].

The artificial intelligence aims at programming intelligence into machines by learning from experiences and adapting to changes in the environment to simulate human decision making and reasoning processes. The machine learning (ML) a part of AI, could be programmed with a set of agricultural data to perform various farm operations without human interference, which enable to solve various complex task[10]. Meanwhile, AI could not work until, a hardware-software interface is embedded, which processes these algorithms and logic based concepts[5]. Thus, automation/ AI in agriculture which is also known as 'digital agriculture' has become the main concern and emerging field globally. AI enables the users to control pesticides application, weed management, irrigation scheduling, and various crop parameters. It has proven emerging technology in smart agriculture that increases the gain from soil precisely and strengthens the soil fertility also[5].

9.2 Role of AI in Water Management in Agriculture

Agriculture sector consumes around 85 per cent of global fresh water. It has been increasing continuously in the wake of population growth. The conventional irrigation method operated manually are not enough to control water losses, thus automated system is introduced into the agriculture sector. Automated system is fed with various agricultural data such as humidity, wind speed, solar radiation, plant growth stages, plant density, and pest behavior[2]. The irrigation is controlled through wireless network by automatic irrigator to ensure the soil fertility and water use efficiency[11]. The smart technologies for irrigation is meant for less human intervention, which automatically sense soil moisture, fertility, and maintains the moisture and nutrient content of the soil accordingly. Micro-controllers are used for ON/OFF the irrigation pump. Arduino and Raspberry pi3 micro-controller have been used in the robots for sensing the soil moisture. The soil moisture is sensed in a regular interval and send the analogue signal to Arduino micro-controller, which converts the analogue signal into digital. Furthermore, the signal is sent to the Raspberry pi3 micro-controller (embedded with KNN algorithms) which sends a signal to Arduino to start the irrigation pump based on the soil moisture content[12]. Such "remote sensors controlled" based robots could increase the crop productivity of around 40%[13]. This sensor senses moisture either by dielectric method or Neutron moderation technique[2]. Likewise, different sensors such as temperature sensor to control temperature, pressure regulator sensor for pressure control, and molecular sensor for crop growth enhancement could be adopted. Zigbee and hotspots are used to process the data through wireless network[14].

The irrigation water applied to the field mainly depends upon farm conditions and farmer's behaviour which ultimately affects the irrigation systems. This could be solved by using a hybrid technique consist of Artificial Neural Network (ANN), Fuzzy Logic, and Genetic Algorithms. Fuzzy logic is an AI technique which is applied as a Fuzzy Interference System (FIS) (Fig. 9.1) to transfer the logics into mathematical and computation structures in estimation of daily crop water demand. However, it works on trial and error method, so finding of actual requirement could not be assured. To overcome the above limitation of FIS, the fuzzy logic and genetic algorithms are combined which is known as "genetic fuzzy system"[15]. The ground water level fluctuation also plays an important role in irrigation scheduling. Adaptive neuro-fuzzy inference system and fuzzy logic techniques are used to predict and simulate the ground water fluctuation[16], where, flexible decision support system helps in optimization of irrigation scheduling. The flexible irrigation decision support system consists of user interface, a knowledge base, and an inference engine. This support system enables the users to obtain suitable water allocation plan for crop on a temporal and spatial fashion using a software interface[17].

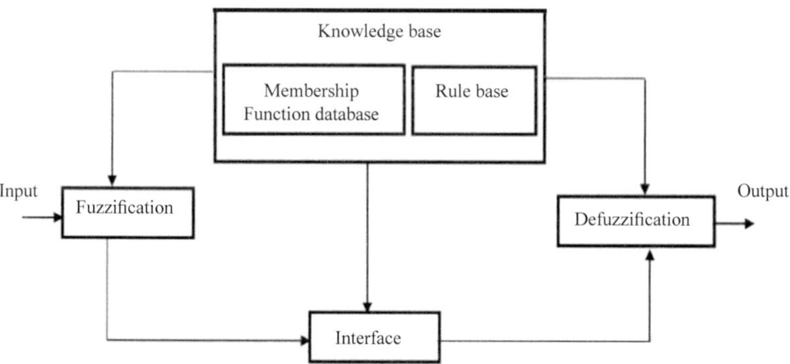

Fig. 9.1: A typical structure of fuzzy interface system

Robotics are playing a vital role in agriculture production and management system. The designing of autonomous agricultural tools has been emphasizing since many decades to improve the overall efficiency of the agricultural machineries and tools. However, these machineries/ tools for conventional farming are not much efficient[18]. Thus, robots are now being emerging into agriculture, which efficiently performed various agricultural operations like weeding, irrigating the land, farm guarding, protecting the crop from adverse climate conditions, and managing the individual plant in various unfamiliar ways[2].

9.3 Plant Growth Observation and Management through AI Embedded System

The observation of plant growth is a complex work, which requires accuracy and expert knowledge. The plant growth parameter such as, plant height, leaf area estimation, leaf chlorophyll level, total biomass production, etc. are some complex task associated with growth dynamics measurement. It is also important to manage plant's long-term (persist up to several days) and short-term (5-6 h) stresses. Thus, modelling of this stage could reduce the plant stresses by predicting and preventing from these stresses. Low power embedded system with AI based on Recurrent Neural Network (RNN) (also known as long short-term memory network) is capable to measure leave growth dynamics *in-situ*. RNN is a class of ANN, where nodes contain the feedback responses and enable the storage of information about their internal state. The state-of-the-art mobile graphical chips are used for smart analysis. The system offers autonomous operation[19]. Image analysis are also embedded to know the plant characteristics. 2-D, 3-D or combination of both approaches are used. Moreover, fruit characteristics could also be analyzed simultaneously with computer vision and machine learning based solution[20]. Despite of 2-D imaging as it requires complicated software, laser scanning is preferred for statistical analysis of leaf area estimation[21].

9.4 Smart Field Operation and Preparation through AI Embedded Agricultural Machineries

The demand in increase in agriculture productivity leads to a proportional increase in energy consumption in farming. Therefore, achieving sustainable agriculture, the relationship between energy consumption and grain yield could be thoroughly studied through automated systems such as ANFIS (Adaptive Neuro Fuzzy Interface System) and ANN (Artificial neural networks) models. ANFIS is bases on FIS (Fuzzy interface system) in which the limit of input variables could be kept five, if it exceeds, then the modelling will show error. Machinery, diesel fuel, human labour, nitrogen, potassium, phosphate, electricity, water for irrigation, FYM, pesticides, and seed are considered as energy inputs for ANFIS model. The input parameters are grouped into different topology to study which combination is most accurate for ANFIS model study. For ANN models, MATLAB (R2012a) is used in which the modifications of hidden layer and 1 to 30 neurons are assessed to generate different results.

From statistical study, data are randomized into three data sets such as training, validation, and testing sets out of which best model is decided on the basis of testing data sets. Root mean square error, mean absolute percentage error, correlation coefficient is taken as the basis of deciding the best model for such study. The ANFIS model is more efficient than ANN model in predicting the wheat grain yield on the basis of energy input with higher R (0.976) and smaller MAPE (0.4) and RMSE (0.046) values where the ANN model has R value of 0.92, higher value of MAPE(0.1) and RMSE (0.9) values[22].

Effectiveness assessment of tractors could be done by applying fuzzy set theory by analysing maintainability, reliability and functionality performances. In this way, the model is useful in concluding any decision regarding purchasing, system maintenance and operation. Fuzzy proposition is used for presenting the statement including linguistic variables (indicated various grades of effectiveness) derived from information available regarding the observed technical systems. Fuzzy composition model illustrates the influence structure of indicators on the effectiveness performance of the systems. Five linguistic variables are taken as the first step for development of fuzzy model, for defining the effectiveness assessment such as poor, adequate, average, good and excellent. Max-min composition as a synthetic model is assessed on the reliability, maintainability, and functionality partial indicators. The opinion of engineers working on the operation and maintenance of tractors are considered to calculate effectiveness of the tractors. This model does not need a complex and expensive infrastructure for effectiveness determination as it requires the expert judgements and estimations. It will be helpful for corrective actions in purchase and construction of equipment, alteration in policy of maintenance[23]. Prediction of tractor repair and maintenance cost is the principal factor in determining in its replacement. Conventional mathematical models are used to derive the relationship between tractor age and repair maintenance cost but associated with major drawbacks of recklessness and uncertainties. To overcome this problem, ANN are applied in predicting repair and maintenance cost of the tractor. The existing data of studied tractors are collected from the company and are split into test and training sets. Training sets are used to estimate the model parameters while test sets are used to inspect the generalization of the proposed model. Cumulative cost index is used to compare repair and maintenance cost of various model tractors despite of different purchase price. Cumulative hours of usage (for tractor age determination) parameter is considered as input variables while cumulative repair cost, oil cost and fuel cost are taken as output variables for ANN model. Multilayer Perceptron (MLP) with Basic Back-propagation (BB) and Back-Propagation with Declining Learning-Rate Factor (BDLRF) is employed to predict the repair and maintenance cost of the tractors with MATLAB software. The developed networks to study the relationship between tractor age and repair maintenance costs are three layered feed forward type using both BB and BDLRF training algorithms. The developed ANN model statistically significant with p values greater than 0.9 implies that it can effectively use in predicting the repair maintenance cost of the tractors. Out of both algorithms used, BDLF is more efficient in providing accurate results under the studied conditions. It eliminates the use of unreliable regression models, as it doesn't consider any fixed dependency between input and output parameters and fast and economic in use[24].

By having the knowledge of working area of the agricultural farm machineries, AI helps in determining the machine effect on the crop yield, operation cost, validation of the machine usage, efficient utilization of resources and the behaviour of the machine operator. The conventional methods such as measurement of working area by tape and GPS system is a very time consuming and inefficient when the fields are irregular in shape. AI modules are used in determining the work area of cultivator, rotavator, and laser-leveller in different regular and irregular field shapes in dry conditions. The IoT (Internet of Things) module along with smartphone and GPS are mounted on the machine in which the sensor data send to the smartphone with addition of latitude and longitudes of GPS into it. The smartphone directs the data into Amazon Web Services (AWS) where AI algorithms are applied into the accumulated data. Field boundary detection of both regular and irregular shape of fields are calculated using Convex Hull with Delaunay Triangulation, Concave Hull using KNN. The working area of field are determined using Delaunay Triangulation, Concave Hull using KNN without FLANN, Concave Hull using KNN with FLANN, Contact-Based. For regular shape field, both algorithms are appropriate while for irregular shape Concave Hull using KNN is well fitted for determining the field boundary conditions. Concave Hull using KNN without FLANN and with FLANN is more efficient in calculating the work area of both regular and irregular shapes in comparison to the Delaunay Triangulation algorithm. The KNN with contact-based algorithm is help in determining the number of field runs and cumulative work done in the field[25].

9.5 AI in Food Processing Industries

Automated computerized process is gaining momentum in food industry as they are highly efficient in online continuous monitoring of quality and different properties of food products. Artificial intelligence such as Computer vision system (CVS), artificial neural network (ANN), sensor technology and fuzzy logic in combination with various drying methods of fruits and vegetables provides a convenient and cost effective online monitoring of drying. Conventional drying methods such as hot air drying, freeze drying are profusely used for drying of fruits and vegetables in food industry but associated with major drawbacks such as damage to colour, flavour, nutrient of products and quite expensive. Empirical mathematical models are used for modelling of drying process but they are not feasible for different range of drying conditions as they are experiment specific. CVS with laser light backscattering assisted hot air drying of papaya has been found to be more appealing in predicting the colour change in comparison to conventional colorimetric method. The multi-regression model using Computer vision system (CVS) with laser light backscattering predict, optimize and control the colour attributes of the papaya products more accurately with $R^2 >0.95$, while colometric method is not feasible for measurement of Chroma values[26] (C*) with R^2 values of 0.45.

Grading of figs or any fruits are such a strenuous and unreliable process due to non-automated method of grading. CVS method being economic and easily operated provides a better grading system based on the colour parameter in comparison to Chroma meter. The browning index and X-colour coordinates are used to perform the test under dynamic conditions while highly speed figs are transported under camera and graded into different classes based on the vision system decision[27].

Temperature and time are used as input parameters to determine the antioxidant capacity and total anthocyanin content of mahaleb puree during infrared drying through the application of nonlinear regressor model ANN, polynomial regressor (PLN) and Least-squares support vector machines (LS-SVM). LS-SVM model is the best for explaining the drying kinetics of mahaleb puree during the infrared drying while PLN and ANN model seems to be incongruous in higher temperature conditions. But ANN model is most suitable for modelling both the drying kinetics and the antioxidant capacity and total anthocyanin content of mahaleb puree during infrared drying[28].

Various mathematical models such as zero order, first order and fractional conversion models are applied in the study of colour changes of different fruits and vegetables however they are suitable for only a specific conditions. In order to overcome these shortcomings, AI models are able to determine the required parameters and model this complex nonlinear system based on the input data where the variables relations are unidentified. BPNN, ELM and BELM artificial intelligence models are used to predict the colour changes of mushroom slices during hot air impingement drying of mushroom slices. The drying time, air temperature and air velocity are used as input variables and colour indexes such as lightness, redness and yellowness taken as output variables. Root mean square error and coefficient of determination was used to determine the significance of the developed BELM model. The number of hidden neurons which influences the precision of these models are taken in the range of 10 to 90 with interval of 5, to obtain the optimal topology of these models which are developed with Mat lab software R2015a. The R^2 value of ELM and BELM is increased while it is drastically decreased for BPNN model until the number of hidden neuron 45, which signifies the former models are more significant. BELM model is time saving and overcomes the problem of over fitting of ELM model. The lowest R^2 (0.9725) and higher RMSE (0.0563) value of BELM model proves that it is an efficient model for predicting the colour changes of agricultural products during drying[29].

ANN and GN model are applied in brewing process as they observed the complex incoherent relationships between output and input parameters in a process. ANN topology is optimized by using a differential evolution algorithm in MATLABR2016a software. The ANN parameters used such as number of neurons and transfer functions from three hidden layers and output layers, training function to be used, number of hidden layers and training and momentum rates are defined as vectors in this study. ANN is formed by randomly formed parameters of each vectors after which mutation step is followed. After mutation, a crossover and selection step is applied to obtain the performance of population vectors on the basis of linear regression coefficient R. ANN topology is optimized by creating a new population vector which continue the algorithm. Multi objective optimizations are not efficiently solved by classical algorithm, so GA is used to overcome this problem. Number of generations, crossover rate, and number of individuals, mutation rate, fitness and migrations are the variables taken for GA. ANN and evolutionary algorithm with a high correlation co-efficient (>0.85) and higher global desirability value (0.78) proved to be an efficient tool to minimize the cost associated with experiment to develop supreme quality beer[30].

Brewing is the use of yeast in the fermentation process of food materials which is quite a complex labour intensive process and difficult to control. AI based software tool are beneficial provides as they provide an effective control and authentically monitor this process online for quick rectification of any faults detected during the brewing process[31].

Grading of egg according to the defects is very laborious and subjected to inaccuracy due to non-automated system of human supervision. In this perspective, AL based automated systems are integrated to resolve this issue and to provide better quality eggs to consumers. The defects such as cracks, blood spots are detected in a machine vision system by using a camera and halogen lamp. Then in pre-processing stage, the acquired image is sent to the MATLAB for further analysis. The images are differentiated into region of interest (egg shell surface) and discarded region (background) to get the colored image of only egg in the segmentation process. Imfill functions are used to discard the undesirable areas that will lead to error in the determination in the size of eggs due to presence of cracks, dirt. Then the number of nonzero pixels of the egg shells are determined and from the correlation between the size of eggs and pixel number, the size of the eggs are determined. Grey scale images and passing of beam light through egg shells are used for crack and broken egg determination respectively. Fuzzy interface system and Simulink model is used for the grading of eggs where input parameters are defects and size of eggs and egg grade is the output parameter. Application of machine vision system with fuzzy tool is more efficient than human grading system as the latter provides more accurate results[32].

9.6 AI in Renewable and Farm Energy Management

Demand of energy in agriculture has been increasing rapidly, meanwhile energy supply remain the key issue in agriculture sector. Thus, energy management in agriculture, and reducing the source of environmental degradation factors need to be addressed through proper energy input-output analysis. The energy use pattern in agriculture could be manage using artificial neural network (ANN), adaptive neuro-fuzzy interface system (AFIS), and multi-layer adaptive neuro-fuzzy interface system[33]. ANN is used to predict energy needs and solving problems in energy related issue. It could also be used for predicting greenhouse gas (GHG) emission in agriculture[34]. Bolandnazar[33] developed a model based on AI to predict the output energy in potato cultivation. The Cobb-Douglas modelling incorporated to estimate the effects of input energy on potato production. The Radial Basis Function (RBF) is used as an active function for ANN, where other models equipped with multi-layer perceptron (MLP) as a class of feedforward ANN. RBF is found to be best model in output energy predication.

AI could play an important role in renewable energy sector as well. The power output from the solar photovoltaic system can be forecasted using AI based technique. The combination of wavelength transform (WT) and AI techniques could be interacted the solar PV with solar radiation and temperature to forecast the power output one-hour ahead. The WT system control the ill-behaved of PV power time-series data wile AI is used to capture the nonlinear fluctuation of PV system[35]. The AI technique could also be used to optimize the hybrid renewable energy system. Genetic algorithms, particle swarm, simulated annealing, and hybrids models are used for the system optimization. These techniques are found to be promising in optimization and modelling of hybrid renewable energy systems. Software tools such as HOMER, HYBRID2, HOGA, OptQuest, LINDO etc. are used along with[36]. The AI technique has also been applied for detection of cracks and damage in blades of wind turbine using AI-based image analytics. Drones are used to capture the images, and deep learning with keras frame work written in python is used to train the ANN model to detect the fault[37].

9.7 Conclusion

Artificial intelligence is an emerging field in agriculture. Automation of agricultural machineries and its key control through mobile application is enabling the users to cultivate crops in a sustainable way. AI as an interface making all the farming activities reliable and sustainable. The increase in crop production, efficient pest control, and automation in irrigation system and fertilizer application without human intervention has brought the farming to its next level. AI is not limited to farming practices only but also offering its benefit to the food process industries in optimization and control of process parameter independently. Observation of plant growth and its nourishing makes the technology more attractive. The application has also been embedded with renewable energy production system. However, it is in its nascent stage, thus, need to be evaluated and applied more in the renewable energy sector. The application of AI in developing country like India has a great scope in the agriculture sector where most of the farming practices are conventional. With the advent of state-of-the art technologies in agriculture for crop cultivation, harvesting, agricultural processing, farm energy production need to be embedded with AI for sustainable agriculture. The use of drones and robots could help in precision farming in the country but at the same time initiative from the government and private firms are required to encourage the farmers. Attractive and reliable policy framework are need of the hours to encourage such technologies in India.

Acknowledgment: Authors are sincerely thanks to Mr.Himanshu Kumar and S. S. Sahoo, CRDT, IIT Delhi for preparing the draft of this chapter.

References

1. Sylvester G, editor. E-agriculture in action: Drones for agriculture. Food and Agriculture Organization of the United Nations and International Telecommunication Union; 2018.
2. Talaviya T, Shah D, Patel N, Yagnik H, Shah M. Implementation of artificial intelligence in agriculture for optimisation of irrigation and application of pesticides and herbicides. Artificial Intelligence in Agriculture 2020; 4: 58-73.
3. Godfray HC, Beddington JR, Crute IR, Haddad L, Lawrence D, Muir JF, Pretty J, Robinson S, Thomas SM, Toulmin C. Food security: the challenge of feeding 9 billion people. Science. 2010; 327(5967): 812-818.
4. Hodges RJ, Buzby JC, Bennett B. Postharvest losses and waste in developed and less developed countries: opportunities to improve resource use. The Journal of Agricultural Science. 2011; 149(S1):37.
5. Jha K, Doshi A, Patel Poojan, Shah M. A comprehensive review on automation in agriculture using artificial intelligence. Artificial Intelligence in Agriculture. 2019; 2: 1-12._
6. Kumar H, Panwar NL._Experimental investigation on energy-efficient twin-mode biomass improved cookstove. SN Applied Science. 2019; 1: 760.

7. Mogili UR, Deepak BB. Review on application of drone systems in precision agriculture. Procedia computer science. 2018; 133: 502-9.
8. Shah G, Shah A, Shah M. Panacea of challenges in real-world application of big data analytics in healthcare sector. Journal of Data, Information and Management. 2019; 1(3-4):107-16
9. Zhao B. The Application of Artificial Intelligence in Agriculture. InJournal of Physics: Conference Series 2020 Jun 1 (Vol. 1574, No. 1, p. 012139). IOP Publishing.
10. Muthukrishnan N, Maleki F, Ovens K, Reinhold C, Forghani B, Forghani Reza. Brief History of Artificial Intelligence. Neuroimaging clinics of North America. 2020; 30 (4): 393-399.
11. Kumar G. Research paper on water irrigation by using wireless sensor network. International Journal of Scientific Research Engineering & Technology (IJSRET). 2014:3-4.
12. Shekhar Y, Dagur E, Mishra S, Sankaranarayanan S. Intelligent IoT based automated irrigation system. International Journal of Applied Engineering Research. 2017; 12(18):7306-20.
13. Savitha M, UmaMaheswari OP. Smart crop field irrigation in IoT architecture using sensors. International Journal of Advanced Research in Computer Science. 2018 ; 9 (1): 302-306.
14. Wartharajalu K, Ramprabu J. Wireless irrigation system via phone call and SMS. International Journal of Engineering and Advance Technology. 2018; 8 (2S): 397-401.
15. Perea RG, Poyato EC, Montesinos P, Díaz JR. Prediction of applied irrigation depths at farm level using artificial intelligence techniques. Agricultural Water Management. 2018; 206: 229-40.
16. Zare M, Koch M. Groundwater level fluctuations simulation and prediction by ANFIS- and hybrid Wavelet-ANFIS/Fuzzy C-Means (FCM) clustering models: Application to the Miandarband plain. Journal of Hydro-environment Research. 2018; 18: 63-76.
17. Yang G, Liu L, Guo P, Li M. A flexible decision support system for irrigation scheduling in an irrigation district in China. Agricultural water management. 2017; 179: 378-89.
18. Dursun M, Ozden S. A wireless application of drip irrigation automation supported by soil moisture sensors. Scientific Research and Essays. 2011; 6(7):1573-82.
19. Shadrin D, Menshchikov A, Somov A, Bornemann G, Hauslage J, Fedorov M. Enabling Precision Agriculture through Embedded Sensing with Artificial Intelligence. IEEE Transactions on Instrumentation and Measurement. 2020; 69(7): 4103-4113.
20. Pouladzadeh P, Shirmohammadi S, Al-Maghrabi R. Measuring calorie and nutrition from food image. IEEE Transactions on Instrumentation and Measurement. 2014; 63(8):1947-56.
21. Yang X, Strahler AH, Schaaf CB, Jupp DL, Yao T, Zhao F, Wang Z, Culvenor DS, Newnham GJ, Lovell JL, Dubayah RO. Three-dimensional forest reconstruction and structural parameter retrievals using a terrestrial full-waveform lidar instrument (Echidna®). Remote sensing of environment. 2013; 135:36-51.
22. Khoshnevisan B, Rafiee S, Omid M, Mousazadeh H. Development of an intelligent system based on ANFIS for predicting wheat grain yield on the basis of energy inputs. Information processing in agriculture. 2014;1(1):14-22.
23. Miodragovic R, Tanasijević M, Mileusnic Z, Jovancic P. Effectiveness assessment of agricultural machinery based on fuzzy sets theory. Expert Systems with Applications. 2012;39(10):8940-6.
24. Rohani A, Abbaspour-Fard MH, Abdolahpour S. Prediction of tractor repair and maintenance costs using Artificial Neural Network. Expert Systems with Applications. 2011;38(7):8999-9007.
25. Waleed M, Um TW, Kamal T, Khan A, Iqbal A. Determining the Precise Work Area of Agriculture Machinery Using Internet of Things and Artificial Intelligence. Applied Sciences. 2020;10(10): 3365.
26. Liu ZL, Nan F, Zheng X, Zielinska M, Duan X, Deng LZ, Wang J, Wu W, Gao ZJ, Xiao HW. Color prediction of mushroom slices during drying using Bayesian extreme learning machine. Drying Technology. 2019; 9:1-3.

27. Benalia S, Cubero S, Prats-Montalbán JM, Bernardi B, Zimbalatti G, Blasco J. Computer vision for automatic quality inspection of dried figs (Ficus carica L.) in real-time. Computers and Electronics in Agriculture. 2016; 120: 17-25.
28. Udomkun P, Nagle M, Argyropoulos D, Wiredu AN, Mahayothee B, Müller J. Computer vision coupled with laser backscattering for non-destructive colour evaluation of papaya during drying. Journal of Food Measurement and Characterization. 2017; 11(4): 2142-50.
29. Isleroglu H, Beyhan S. Intelligent models based nonlinear modeling for infrared drying of mahaleb puree. Journal of Food Process Engineering. 2018; 41(8): e12912.
30. Takahashi MB, Coelho de Oliveira H, Fernández Núñez EG, Rocha JC. Brewing process optimization by artificial neural network and evolutionary algorithm approach. Journal of Food Process Engineering. 2019; 42(5): e13103.
31. Vassileva S, Mileva S. AI-based software tools for beer brewing monitoring and control. Biotechnology & Biotechnological Equipment. 2010; 24(3): 1936-9.
32. Omid M, Soltani M, Dehrouyeh MH, Mohtasebi SS, Ahmadi H. An expert egg grading system based on machine vision and artificial intelligence techniques. Journal of food engineering. 2013; 118(1): 70-7.
33. Bolandnazar E, Rohani A, Taki M. Energy consumption forecasting in agriculture by artificial intelligence and mathematical models. Energy Sources, Part A: Recovery, Utilization, and Environmental Effects. 2020; 42(13): 1618-1632.
34. Khoshnevisan B, Rafiee S, Omid M, Mousazadeh H. Prediction of potato yield based on energy inputs using multi-layer adaptive neuro-fuzzy inference system. Measurement. 2014; 47: 521-30.
35. Mandal P, Madhirab STW, Haque AUl, Mengc J, Pineda RL. Forecasting Power Output of Solar Photovoltaic System Using Wavelet Transform and Artificial Intelligence Techniques. Procedia Computer Sicence. 2012; 12: 332-337.
36. Zahraee SM, Assadi MK, Saidur R. Application of Artificial Intelligence Methods for Hybrid Energy System Optimization. Renewable and Sustainable Energy Reviews. 2016; 66: 617-630.
37. Reddy A, Indragandhi V, Ravi L, Subramaniyaswamy V. Detection of Cracks and damage in wind turbine blades using artificial intelligence-based image analytics. Measurement. 2019; 147: 106823.

10

Robotics in Agriculture

10.1 Introduction

The agriculture sector is also transforming through technological intervention and advancements drastically to improve several processes and crop and livestock production systems. These advancements have resulted in minimizing operation and production costs, reduction of environmental impact and optimization overall production cycle. The major issue in the agriculture sector is human-intensive operation and non-availability of sufficient labor on time. The on-field task like fruit harvesting, intra-row weed control which is supposed to be more difficult to perform by the traditional machinery and manually. This has made to think to introduce the autonomous tractors and robotics platform in field operation[1]. Timely activities and treatment play an important role in agriculture and agriculture production. The farmers generally face many problems in performing the task on the field due to various limitations. The cost is one of the main parameters which need to keep the farmer in a competitive market and reduce the cost of operation as low as possible. Also, the growing population demand more and more food which put pressure on the agriculture sector to raise crop production. In this situation, agriculture robots can play an important role to address the challenges in the farms. A robot can professionally perform the task and can automate the tasks which are difficult for manual labor. These are used in the agricultural lands and they are also completely autonomous. These robots are used in several applications in agriculture and can also work autonomously.

The robotic system has been widely used in industrial production and warehouses in an almost controlled environment[2]. In the early 1960s, the agriculture and forestry sector initiated research into driverless vehicles with a focus on automatic steered system and autonomous tractors[3]. Robotics has made an impact on agriculture and forestry since past few years. The purpose of minimizing the damage to the growing zone of soil tends to cater to the application of automatic vehicle guidance. The interest in the development of robotics system in agriculture has led many experts to explore the possibilities to develop more rational and adaptable vehicles based on a behavioural approach. The combination of the sensor system, Communication Technologies, Positioning System (GPS) and Geographical Information System (GIS) have made the way in the development of new autonomous vehicles which can be applied for a high-value crop in the agriculture and horticulture sector as well as landscape management[2]. Specialized sensors (machine vision, GPS), real-time kinematics (RTK), laser-based equipment, and inertial devices, actuators) hydraulic cylinders, linear and rotational electrical motors) and electronic equipment (embedded computers, industrial PC and PLC) have enabled the integration of many autonomous vehicles, particularly agriculture robots[4-7]. It can be seen that the autonomous and semiautonomous system equipped with the proper implements provide accurate positioning and guidance in the working field which makes them capable of conducting precision agriculture task. The implements (variable application rate of fertilizer or spray, mechanical intra-row weed control, seed planter) can be automated with the same type of sensors and actuators[8-10]. The vehicle integration with particular implements a central external computer must be used to coordinate the arrangement. To make the commercially reliable, efficient and cost-competitive agriculture machinery, vehicle implement should use the minimum hardware[11] As an example GPS enabled tractor to follow the row of crops, performing headland turn at the end of a row can make the substantial transformation in tractor application in agriculture. Color sorting and grading of produce though involve sensing technology, it needs measures of produce handling which also related to robotics[12]. The manufacturing industry was the first to witness the introduction of the Robotics under the name Robots for serial manipulation. The core of robotics involves manipulation and related kinematics which later transforms into intelligent automation. However, in agricultural practices, sensing is much more emphasises rather than manipulation[12].

Due to the introduction of agricultural robots, there will be replacement of the heavy and large tractors. This will result in non-compaction of soil, more attention to the plants and reduced cost of operation. The light, slow and small robots aided with GPS, smartphones can work autonomously without the requirement of manpower.

10.2 Agriculture Robotics

The manual labor outflow from agriculture and focus on mechanization has created a great opportunity for agriculture robotics[1]. The agriculture robotics technologies application is not just limited to the field of agriculture but also look at the challenges in agriculture to develop new techniques and systems. Billingsley et al have broadly given the scope of agriculture robotics in the field of following fields[13]:

(i) Farming environment for produce, which presents large, semi-structured open spaces

(ii) Farm facilities for a variety of animals, which could also present large semi-structured open space (e.g cattle grazing areas) or environments that are more built up, such as caged chicken farms

(iii) Forestry

(iv) Aquaculture

As the agricultural sector faces a shortage of human labor, agriculture robots can work autonomously and intelligently. Automatic agriculture vehicle can be grouped into four categories as[10,14].

(i) Guidance (navigation of vehicle within the agriculture environment)

(ii) Detection (extraction of biological features from the environment)

(iii) Action (execution of a task for which the vehicle was designed)

(iv) Mapping (construction of a map of the agricultural field with relevant features)

Agricultural robotics technology in terms of perception, manipulation and autonomous vehicles have improved drastically. Perception is achieved through the variety of sensors, which can be used even in a small focused workspace or monitoring vast agricultural facilities through distributed sensors. Manipulation tends to simplify the task with fewer uncertainties or lower accuracies such as spraying. The complicated task with high variability like picking tomatoes and fruits growing on different attitude where there may have obstacles likes branches and leaves also. Autonomous vehicles have gained much interest in agriculture robotics technology which showes the possibility of operating the many unmanned vehicles in a coordinated and robust manner to operate vast farming facilities[15].

Autonomous agricultural application has to take two operational subsystems i.e vehicle and (ii) implement and put them to work together.

(i) Vehicle: The vehicle is the module which controls the behaviour of implement (absolute position and orientation) and must adapt both the type of crop and operation on the crops. The vehicle generally carries or tow the implements providing the necessary energy to the implements. Adapting commercial tractor is easy to configure autonomous vehicle is easier and more efficient.

(ii) Implement: The implement is a device designed for performance on the crop, such as herbicide and pesticide booms and mechanical and thermal weed remover. PLCs and computers are used to control individual elements like nozzles and burners used.

A collaborative aerial and ground-based robotic system provides significant benefits to much practical application, as they can combine the advantages of multiple heterogeneous platforms[16] which has more efficient application in precision agriculture like mapping weed distribution or crop nutrition status indicators as shown in Fig.10.1. It shows that an unmanned aerial vehicle (UAV) continuously survey the field over the growing season (top left), collects the data about crop density and weed pressure (top right) and coordinates and share information unmanned ground vehicle (UGV) (bottom left) that is used for targeted intervention and data analysis (bottom right). The gathered and merged information is then delivered to form operation for high-level decision making[16].

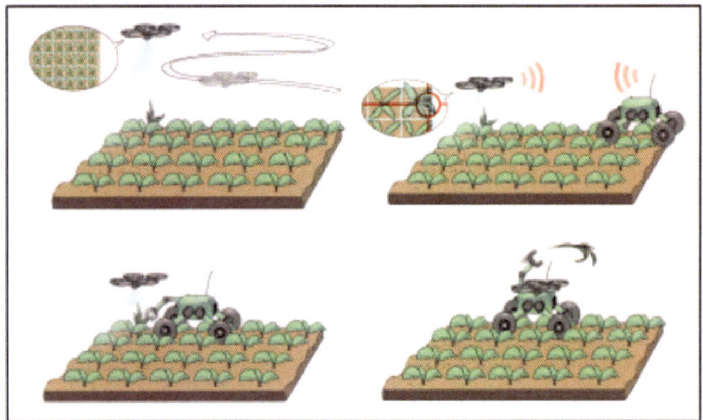

Fig. 10.1: A conceptual overview of the aerial and ground-based system

10.3 History of Agriculture Robots

Robotics development in agriculture was initially taken up during early 1920. The research on providing automatic guidance to the vehicle which given rising to development between 1950 and 1960. As vehicle needed the cable system for their guidance along the path it travels, the concept was not much success but the development of agricultural robots continued as technology in various sectors. The development of computers in the 1980s, made the guidance for the vision of the machine possible and successive development in the following years for the harvesting of oranges with the help of agriculture robots in countries like USA and France. However, there was a concern of the use of those robots due to security reason, difficulty in picking crops and varying conditions of the environment.

10.4 Agriculture Robots and Types

The robotic system in agriculture field operation are classified according to major field operation as (i) weeding, (ii) seeding, (iii) disease and insect detection, (iv) crop scouting (plant monitoring and phenotyping), (v) spraying, (vi) harvesting, (vii) plant management robots, and (viii) multipurpose robotic system[17]. The agriculture robots are mean to perform the task professionally which are supposed to be difficult manually on the farm. Following are the different types of agriculture robots which are used in farm operations

a) **Precision Agriculture:** The robot application is for small farms and vineyards which follows the precision agriculture techniques. These robots monitor the activities like the respiration of soil, photosynthesis activity and biological aspects.

b) **Monitoring the pollution**: These type of robots monitors the pollutions like carbon dioxide emission and nitrous oxide mainly from ground level to help farmers to reduce the emissions

c) **Control of weeds:** Specially designed agriculture robots for controlling weed, navigate the weed scan autonomously and spray the herbicide on the targeted points to eliminate or destroy the weeds. This approach helps to avoid the exposure of crop to herbicide and prevent the growth of weeds which are resistant to the herbicide

d) **Livestock Ranching:** These robots are used to herd the livestock in huge ranches, monitors the animals in the livestock. These robots also make sure about the health of animals and the sufficient grazing space is provided.

e) **Automation of nursery:** These robots mainly used in nurseries to move the plants around the greenhouse efficiently and address the shortage of labour also.

f) **Harvesting of the crop:** To harvest the crops, these agriculture robots are specialized in working round the clock for fast harvesting operation. These robots efficiently perform the task of 25 to 30 workers.

g) **Harvesting of fruits:** The harvesting of fruits is performed using these type of agriculture robots which are designed to identify the fruits and pick them without causing any damage.

h) **Seeding and planting:** These robots are equipped with a three-dimensional vision system which makes them able to plant and seed the crops for good growth with accuracy. They are used in the farming of lettuce and vineyards.

10.5 Applications of Robotics in Agriculture[18]

Agriculture robots are designed to perform the specialized task in agriculture as shown in Fig. 10. 2. The major field operation is as follows:

10.5.1 Monitoring the Crop and Analysis

Monitoring the crops in huge fields make it difficult for manual labour. The technological advancement like geo-mapping and sensors can help farmers to get a high level of information about crops growing in the field. The drones and robots which are at the ground autonomously collect the data and task. Some of the companies offer the hardware and analysis software which help the farmers to use the technology with the help of a smartphone and get the real-time data and analysis of grown crops. The ground robots provide the farmers can get very close to crops.

10.5.2 Fertilization and Irrigation

The traditional method used a huge amount of water inefficiently for irrigation and fertilization of crops. Precision irrigation assisted by the robots will help to decrease the water requirement by targeted irrigation to needful plants. Ground-based robots autonomously navigate among the crop rows and provide water directly at the base of every plant. The robots provide an advantage to access the areas where the other machines of agriculture could not go and also can reliably manage to timely irrigation.

10.5.3 Crop Weeding

Traditional weeding is more time consuming, laborious and repetitive agriculture activity which almost consume 40 per cent of the labor effort in weed management in developing countries and small farm holders[19]. The traditional way of spraying pesticide and weedicide onto the agricultural field causes harm to the environment. Agriculture robots can perform this task efficiently by adapting micro-spraying method which helps to decrease the quantity of herbicide requirement for growing crops. The technology aid from computer vision to detect the weed and spray the targeted amount of herbicide onto the crop. Some robots perform weeding by using computer vision to detect plants which need to be moved by tractor to uproot it. Mechanical weeding and chemical weeding are two approaches which have been adapted. Mechanical weeding robots have row detection precision[20-22] form 6 mm to 25 mm. The weeding efficiency[20,23] i.e. performance for weed removal can give as high as 90 %. Some of the robotic system used for weeding have the characteristics and performance as shown in Table 10.1.

Table 10.1: Weeding agriculture robotic systems characteristics for various crops

Crop	Perception Sensors	Weed Detection	Weed Control
Maize[24]	Cameras, optical and acoustic distance sensors	Yes	Chemical
Carrot[25]	RGB infrared camera	Partly	Chemical
Sugar beet[20]	Color camera	Yes	Mechanical
Rice[21]	Laser range finder, IMU	No	Mechanical
Beetroot[25]	Color camera, artificial vision, compass	Yes	Chemical
Grapes[27]	IMU, hall sensors, electromechanical sensor, a sonar sensor	No	Mechanical
Tomato[28]	Color camera, Sensor watch	Partly	Chemical

10.5.4 Thinning and Pruning

Thinning involves decreasing the plant density by removing excess plants to provide the other plant with a better chance to grow. The robots will decide which plant to remove and which to keep by computer vision. Likewise pruning which is tough task is done by robots by cutting off the plants parts to improve the growing condition.

10.5.5 Phenotyping

Phenotyping robots use vision systems to gather through them as much as information as possible about plant/ fruit under inspection. These type of robots focus mostly on morphology, plant growth and traits that are related to crop growth. Multiple sensors are used in phenotyping robots. Fully developed phenotype robot[29] for sorghum crop shows detection up to 96 per cent. The task like light penetration, leaf erectness, leaf necrosis and Green Red Vegetation Index (GRVI) can also be measure d by these robots[22].

10.5.6 Autonomous Tractors

Most of the agricultural robots would be attached or hitched to the tractors which are generally driven by the man and robots are designed to adapt the speed of the tractor. However, the autonomous tractor is as mentioned earlier operates could be operated using smartphones etc.

10.5.7 Harvesting and Picking

Harvesting is a major task involved in crop production which requires huge labor work in agriculture. The robotic harvester can be two types: Bulk harvester (one which harvests all fruits/ vegetables) and another is selective harvester (only ripe/ready fruit). The care must be taken to harvest the crop within time and without damaging crop and plant[22]. Harvesting the crops like corn, wheat and barley are an easier task which can be done by a combine harvester and can be automated. Crops like soft fruits are difficult to harvest and required manual skill. Few developed robots can perform harvesting of apples, sweet paper etc. A high-value crop like strawberry suffers from high production cost due to the high labor requirement[30,31]. The harvesting robots have registered the fastest picking time between 7.5 to 8.6 second per strawberry[32,33]. Vegetable crops have tremendous scope to use the robotic system. A cucumber robot claimed picking rate of 80 per cent at an average picking speed of 45 seconds per cucumber[34]. Whereas for eggplant, the picking rate was about 62%, while it needed 64 seconds to harvest an eggplant[35]. Table 10.2 shows the features of a few harvesting robots for different crops.

Table 10.2: Harvesting agriculture robotic systems features for various crops

Crop	Perception	Fastest Picking Speed	Highest Picking Rate
Apple tree[36]	Color camera, time of flight based three dimensional camera	7.5 sec/fruit	84%
Apple tree[37]	Color CCD (Charge Coupled Device) camera, laser range sensor	7.1 sec/fruit	89%
Mushroom[38]	Laser sensor, vision sensor	6.7 sec/mushroom	69%
Strawberry[31]	Sonar camera sensor, binocular camera	31.3 sec/fruit	86%

10.5.8 Nursery

Automation of the nursery has seen increasing trend in agriculture robots as vertical and urbaan farming are becoming more popular. Plant nurseries have the crucial part of material handling. Specially designed robots for unstructured environment (e.g. greenhouse, hoop house etc) are available which work fully autonomous mode. These robots can work safely alongside people and minimal training is required to operate. This helps in reducing the labors and cost, efficient resource management and timely production and inventory control. Nursery bots uses wheels, gripper arms, trays and sensors to indentify and move the plants to desired location[39]. The robots developed specially for nursery (BigTop) uses sensors to observe surrounding environment which enable it to make decision about its behaviour and pass the information to the operator. A pre-programmed algorithm make it work on autonomous mode and perform the required job as per instructions provided by the nursery or agriculture business[40].

10.5.9 Milking in Dairy Farming

There is already good progress in robots used in milking in the dairy farms. Robotic milking reduces the demand for excess labor on dairy farms of all sizes and make farm life more flexible and increase production per cow[41,42]. Some robots used can spray the disinfectant on the udders of the cow while preparing it for milking.

10.5.10 Herding and Shepherding

Some of the robots like drones in countries like New Zealand and Ireland are used for farming sheep and cattle over long and tough terrain. These robots also can assist work on the farms. The robot system allows the nature of the behaviour remotely and the rest of the things taken care of by the robot itself without additional supervision. The ability of these robots to walk the terrain with agility or in some cases uses wheels. These robots can be equipped with a series of technologies according to the need of the site[43].

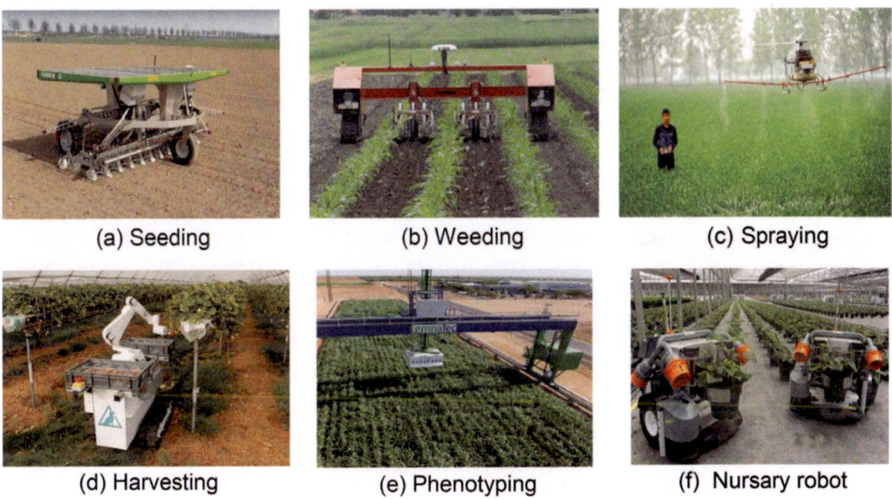

Fig.10.2: Agriculture robots used for different applications

References

1. Marinoudi V, Sørensen CG, Pearson S, Bochtis D. Robotics and labour in agriculture. A context consideration. Biosystems Engineering. 2019;184:111-21.
2. Pedersen SM, Fountas S, Have H, Blackmore BS. Agricultural robots—system analysis and economic feasibility. Precision agriculture. 2006; 7(4): 295-308.
3. Wilson JN. Guidance of agricultural vehicles—a historical perspective. Computers and electronics in agriculture. 2000; 25(1-2): 3-9.
4. Åstrand B, Baerveldt AJ. An agricultural mobile robot with vision-based perception for mechanical weed control. Autonomous robots. 2002;13(1): 21-35.

5. Bakker T, van Asselt K, Bontsema J, Müller J, van Straten G. Autonomous navigation using a robot platform in a sugar beet field. Biosystems Engineering. 2011; 109(4): 357-68.
6. Kalogirou S. The potential of solar industrial process heat applications. Applied Energy. 2003; 76(4): 337-61.
7. Li M, Imou K, Wakabayashi K, Yokoyama S. Review of research on agricultural vehicle autonomous guidance. International Journal of Agricultural and Biological Engineering. 2009; 2(3): 1-6.
8. Gan-Mor S, Clark RL, Upchurch BL. Implement lateral position accuracy under RTK-GPS tractor guidance. Computers and Electronics in Agriculture. 2007; 59(1-2): 31-8.
9. Nieuwenhuizen AT, Hofstee JW, Van Henten EJ. Performance evaluation of an automated detection and control system for volunteer potatoes in sugar beet fields. Biosystems Engineering. 2010; 107(1): 46-53.
10 Pérez-Ruiz M, Slaughter DC, Gliever CJ, Upadhyaya SK. Automatic GPS-based intra-row weed knife control system for transplanted row crops. Computers and Electronics in Agriculture. 2012; 80: 41-9..
11. Rovira-Más F. Sensor architecture and task classification for agricultural vehicles and environments. Sensors. 2010; 10(12): 11226-47.
12. Billingsley J, Visala A, Dunn M (2008) Chapter 46 Robotics in Agriculture and Forestry. Springer Handb Robot
13. Billingsley J, Oetomo D, Reid J. Agricultural robotics [TC spotlight]. IEEE Robotics & Automation Magazine. 2009; 16(4):16
14. Foglia MM, Reina G. Agricultural robot for radicchio harvesting. Journal of Field Robotics. 2006; 23(6-7): 363-77.
15. John Billingsley and John Reid (2009) Agricultural Robotics. IEEE Robot. Autom. Mag.2009: 16–19
16. Pretto A, Aravecchia S, Burgard W. Building an Aerial-Ground Robotics System for Precision Farming: An Adaptable Solution. IEEE Robot Autom Mag. 2020: https://doi.org/10.1109/MRA.2020.3012492
17. Blackmore BS, Fountas S, Gemtos TA, Griepentrog HW. A specification for an autonomous crop production mechanization system. InInternational Symposium on Application of Precision Agriculture for Fruits and Vegetables 2008; 824: 201-216.
18. https://www.agrifarming.in/ (accessed on Nov. 09, 2020)
19. Labrada R. FAO Training on Weed Management. InProc. of the 18th Asian-Pacific Weed Sci. Soc., Conf. May 2001; 1-8.
20. Bakker T, van Asselt K, Bontsema J, Müller J, van Straten G. An autonomous weeding robot for organic farming. InField and Service Robotics 2006 (pp. 579-590). Springer, Berlin, Heidelberg.
21. Kim GH, Kim SC, Hong YK, Han KS, Lee SG. A robot platform for unmanned weeding in a paddy field using sensor fusion. In2012 IEEE International Conference on Automation Science and Engineering (CASE) 2012 Aug 20 (pp. 904-907). IEEE.
22. Fountas S, Mylonas N, Malounas I, Rodias E, Hellmann Santos C, Pekkeriet E. Agricultural Robotics for Field Operations. Sensors. 2020; 20(9): 2672.
23. https://www.naio-technologies.com/en/agricultural-equipment/weeding-robot-oz/ (accessed on Nov. 09, 2020)
24. Klose R, Thiel M, Ruckelshausen A, Marquering J. Weedy - a sensor fusion based autonomous field robot for selective weed control. In: Conference: Agricultural Engineering - Land-Technik 2008: Landtechnik regional und International
25. Utstumo T, Urdal F, Brevik A, Dørum J, Netland J, Overskeid Ø, Berge TW, Gravdahl JT. Robotic in-row weed control in vegetables. Computers and electronics in agriculture. 2018; 154: 36-45.

26. https://www.ecorobotix.com/en/ (accessed on Nov. 09, 2020)
27. Reiser D, Sehsah ES, Bumann O, Morhard J, Griepentrog HW. Development of an autonomous electric robot implement for intra-row weeding in vineyards. Agriculture. 2019; 9(1): 18.
28. Lee WS, Slaughter DC, Giles DK. Robotic weed control system for tomatoes. Precision Agriculture. 1999; 1(1): 95-113.
29. Mueller-Sim T, Jenkins M, Abel J, Kantor G. The Robotanist: a ground-based agricultural robot for high-throughput crop phenotyping. In2017 IEEE International Conference on Robotics and Automation (ICRA) 2017 May 29 (pp. 3634-3639). IEEE.
30. Feng Q, Zheng W, Qiu Q, et al. Study on strawberry robotic harvesting system. In: CSAE 2012 - Proceedings, 2012 IEEE International Conference on Computer Science and Automation Engineering
31. Feng Q, Wang X, Zheng W, Qiu Q, Jiang K. New strawberry harvesting robot for elevated-trough culture. International Journal of Agricultural and Biological Engineering. 2012; 5(2): 1-8.
32. Xiong Y, Peng C, Grimstad L, From PJ, Isler V. Development and field evaluation of a strawberry harvesting robot with a cable-driven gripper. Computers and electronics in agriculture. 2019;157:392-402.
33. Hayashi S, Yamamoto S, Tsubota S, Ochiai Y, Kobayashi K, Kamata J, Kurita M, Inazumi H, Peter R. Automation technologies for strawberry harvesting and packing operations in japan 1. Journal of Berry Research. 2014; 4(1): 19-27.
34. Van Henten EJ, Hemming J, Van Tuijl BA, Kornet JG, Meuleman J, Bontsema J, Van Os EA. An autonomous robot for harvesting cucumbers in greenhouses. Autonomous robots. 2002; 13(3): 241-58.
35. Hayashi S, Ganno K, Ishii Y, Tanaka I. Robotic harvesting system for eggplants. Japan Agricultural Research Quarterly: JARQ. 2002; 36(3): 163-8.
36. Silwal A, Davidson J, Karkee M, et al. Effort towards robotic apple harvesting in Washington State. In: American Society of Agricultural and Biological Engineers Annual International Meeting, ASABE 2016
37. Bulanon D, Kataoka T. Fruit detection system and an end effector for robotic harvesting of Fuji apples. Agric Eng Int CIGR J 2010; 12: 203–2010
38. Rowley JH. Developing flexible automation for mushroom harvesting (Agaricus bisporus). University of Warwick. 2009
39. https://www.public.harvestai.com/ (accessed on Nov. 05, 2020)
40. https://www.ai-systems.ca/bigtop-autonomous-mobile-robot/ (accessed on Nov. 05, 2020)
41. Rodenburg J. Robotic milking: Technology, farm design, and effects on work flow. Journal of Dairy Science. 2017;100(9): 7729-38.
42. Koning K de. Automatic Milking - Common Practice on Dairy Farms. First North Am Conf Precis Dairy Manag 2010
43. https://www.inceptivemind.com/boston-dynamics-rocos-spot-robot-shepherd-new-zealand/13324/ (accessed on Nov. 05, 2020)